高等院校土建学科双语教材（中英文对照）

◆ 建筑学专业 ◆

BASICS

设 计 概 念
DESIGN IDEAS

[德] 贝尔特·比勒费尔德
[西] 塞巴斯蒂安·埃尔库里 编著

张路峰 译

中国建筑工业出版社

著作权合同登记图字：01-2009-6386号

图书在版编目（CIP）数据

设计概念/（德）比勒费尔德，（西）埃尔库里编著；张路峰译.
北京：中国建筑工业出版社，2011
高等院校土建学科双语教材（中英文对照）◆建筑学专业◆
ISBN 978-7-112-11586-0

Ⅰ.设… Ⅱ.①比…②埃…③张… Ⅲ.设计学－高等学校－教材－汉、英 Ⅳ.TB21

中国版本图书馆CIP数据核字（2009）第209820号

Basics: Design Ideas/Bert Bielefeld, Sebastian El Khouli (Ed.)
Copyright © 2007 Birkhäuser Verlag AG (Verlag für Architektur), P.O. Box 133, 4010 Basel, Switzerland
Chinese Translation Copyright © 2011 China Architecture & Building Press
All rights reserved.
本书经 Birkhäuser Verlag AG 出版社授权我社翻译出版

责任编辑：孙　炼
责任设计：郑秋菊
责任校对：关　健　赵　颖

高等院校土建学科双语教材（中英文对照）
◆建筑学专业◆
设计概念
[德] 贝尔特·比勒费尔德　编著
[西] 塞巴斯蒂安·埃尔库里
　　张路峰　译

*

中国建筑工业出版社出版、发行（北京西郊百万庄）
各地新华书店、建筑书店经销
北京嘉泰利德公司制版
北京密东印刷有限公司印刷

*

开本：880×1230毫米　1/32　印张：4¾　字数：140千字
2011年5月第一版　　2011年5月第一次印刷
定价：**18.00**元
ISBN 978-7-112-11586-0
　　（20273）

版权所有　翻印必究
如有印装质量问题，可寄本社退换
（邮政编码100037）

中文部分目录

\\ 序　5

\\ 导言　85

\\ 设计基础　87

\\ 设计与文脉　89
　　\\ 景观与气候　93
　　\\ 城市规划与建筑文脉　95
　　\\ 社会与文化因素　105

\\ 设计与功能　111
　　\\ 满足使用者需求　112
　　\\ 空间布局与内部组织　113

\\ 设计的构成　118
　　\\ 秩序与比例　118
　　\\ 设计的基本元素　122
　　\\ 空间与实体　127

\\ 设计中材料与结构的应用　133
　　\\ 材料和结构作为设计元素　133
　　\\ 对材料的感知　136

\\ 设计概念的形成　138
　　\\ 创造性与创新技巧　140
　　\\ 方法与策略　143

\\ 结语　147

\\ 附录　148
　　\\ 参考文献　148
　　\\ 图片来源　149

CONTENTS

\\Foreword _7

\\Introduction _9

\\Design basics _11

\\Designing in context _15
    \\Landscape and climate _18
    \\Urban planning and architectural context _20
    \\Social and socio-cultural factors _32

\\Design and function _39
    \\Responding to user needs _39
    \\Spatial allocation plan and internal organization _41

\\Constituents of design _47
    \\Order and proportion _47
    \\Basic elements of design _50
    \\Space and bodies _56

\\Working with materials and structures _63
    \\Materials and structures as design elements _63
    \\How we perceive materials _67

\\Arriving at ideas _69
    \\Creativity and creativity techniques_72
    \\Methods and strategies _74

\\In conclusion _79

\\Appendix _81
    \\Literature _81
    \\Picture credits _82

序

设计是一个很难被系统化或类型化的过程。设计方案是尝试不同途径、分析各种影响的"试错"(trial-and-error)过程的产物,特别是对于初学者,当他们在创意生成与设计领域迈开第一步时更是如此。他们往往有了一点想法,却又发现这些想法很难发展下去,而那些想法经常会引出其他的更有意思的想法。这种灵感与愉悦交织、退却与挫折相伴的过程使他们对设计任务的理解逐渐明确,从而使设计方案得以成型。即使对于那些有大量专业知识和多年工作经验的建筑师,设计过程也大致如此。

任何设计都始于对设计概念的追求,即对如何解决设计问题的直觉理解。设计概念是一个漫长旅途的起点,设计师在设计过程中对设计概念进行完善和修正,添加细节,不断地对设计结果进行反思。本书将以此为主题,对设计过程的初始阶段进行探讨,这一阶段对于设计过程和结果的影响和作用是决定性的。本书的目的不仅仅是对各种有效的方法和灵感的来源进行描述,而且要对创造性进行解密。本书的内容旨在鼓励初学者更深一步地去探讨个别的题目与观念。本书的重点并不在于某种特殊的建筑样式或教条化的原理,我们所关注的是一个既简单又复杂的问题:我怎样才能提出一个初始的设计概念?

编者:贝尔特·比勒费尔德(Bert Bielefeld)

FOREWORD

Design is a process that is hard to systematize or typologize. Designs are the result of different approaches, influences and a trial-and-error process, especially when students are taking their first steps in the world of concept generation and design. They try something out and often discover that its potential is limited, yet their idea usually leads to new alternatives and interesting paths. This combination of inspiration and joy, of setbacks and frustration, sharpens their understanding of the design assignment, and a design finally takes shape. Even architects who have been working in the profession for a long time and have a great deal of knowledge experience the design process in this way.

Every design begins with a search for an idea or for an intuitive understanding of how an assignment should be solved. This design idea is the start of a long journey on which the designer defines the idea more precisely, modifies it, adds details and repeatedly rejects results. The current book, *Design Ideas*, is confined thematically to the start of this process, which influences and sets the direction of both the path and, often, the results. Its goal is not only to depict a variety of effective approaches and sources of inspiration, but also to show ways to unlock creativity. The contents are meant to encourage students to explore individual topics and concepts in greater depth. The focus has deliberately been shifted away from specific architectural styles and dogmatic principles. What is at issue here is a simple, yet complex, question: how do I come up with an initial idea?

Bert Bielefeld
Editor

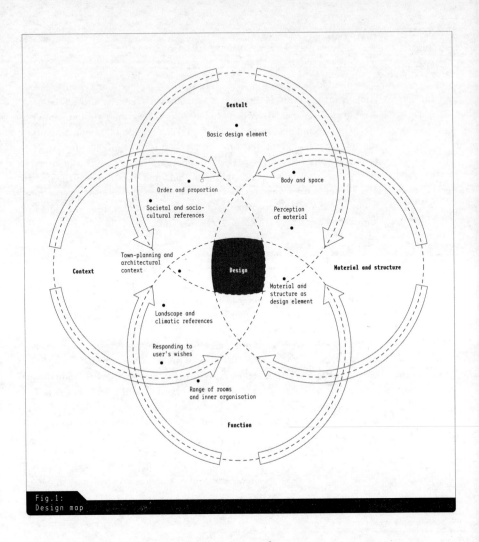

Fig.1:
Design map

INTRODUCTION

Architecture is not created in a vacuum. It is usually a response to the context in which it becomes constructed reality. Architecture is also expected to perform functions, to provide a concrete solution to an assignment, and to come to life through its design and materials. This is why the parameters set out in the "design map" – context, function, form, materials and structure – are directly related to every architectural design. › Fig. 1 They are also the elements in every design for a building. Furthermore, they hold the most potential when it comes to strategies for developing a design idea.

The following chapters will systematically present the parameters relevant to a design and analyse them with an eye to possible design approaches and concepts. Diverse links to other design-relevant factors will be mapped out to underscore possible points of contact and dependencies. These cross-references illustrate the way individual themes are intertwined and should keep you from getting caught in a dead end in the idea phase. In addition, the chapters will include references to exemplary buildings and more advanced texts on architecture so that you can carry out more detailed studies of the discussed methodologies and their architectural application.

These design parameters form a framework you can use when trying to generate your own ideas. They enable you to tap relevant sources of both information and inspiration in a structured way for your initial design steps. At the very start of the design process, it is often helpful to compile all the known information, conditions and perceptions and to visualize them in a consistent fashion. This exercise often reveals unnoticed connections and focuses, while pointing to existing gaps in knowledge and possible contradictions.

The final section of this book introduces different methods and exercises to help you take the first, and often difficult, "plunge" into the design process. It concentrates on the individual points of contact to your design work.

DESIGN BASICS

The design process

Design is a complex, often contradictory, non-linear process. This applies as much to the work of experienced architects as to that of novices, since it is the very nature of the beast. Even if the details of the assignment are clear, the goal of the process is unknown. Learning to design involves embarking on a quest for the methods that make it possible to recognize points of contact and dependencies and to understand the reference system of any given assignment. Architects then adapt these factors to architecture using their knowledge, experience, spatial imagination and creativity.

Every design poses new questions that give you the opportunity to gain fresh knowledge and to create a prototype tailored to the assignment. Designing is not only a central element that links everyone in the architectural profession. It is one of the most interesting aspects of the work.

Questions instead of answers

When a new assignment begins, it is more important to ask the right questions than to embark on a hasty search for simple answers that might not do justice to the assignment's complexity. A large number of these questions will emerge from the immediate context of the assignment. An intense examination of the specific conditions of the design or of exemplary works of architecture can therefore be a promising way to approach the assignment. You can choose from a variety of strategies and methods.

Analysis and inspiration

A common method involves the detailed study and analysis of the most important parameters:

_ Urban planning context/landscape context
_ History of the site
_ User/utilization requirements
_ Other buildings in similar contexts with similar functions

Linking this information with the results of analyses will help you generate ideas that lead to a concrete design concept. In addition to doing scientific analyses, you can pursue other, more playful methods that offer greater freedom because they entail fewer constraints. › Chapter Arriving at ideas

Another approach involves searching for inspiration or an idea at the very start of the process. The idea can be derived from the individual details of the assignment, its requirements, or even from sources of inspiration that bear no direct relation to the assignment. › Chapter Arriving at ideas, Methods and strategies As the work continues, other requirements and levels of design are gradually integrated into the concept. As a result, the design evolves in an ongoing transformation process.

The choice of the right method depends on a person's working style, skills and the concrete assignment. It can differ from design to design. All students should take advantage of the opportunity to try out different approaches and solutions in the course of their studies. The goal is to recognize the strengths and weakness of their individual approaches and to find out which approach suits them best.

Experiences

Personal experiences and perceptions are decisive in the process of generating ideas. With every exercise, you will hone the tools of the design trade and develop a feel for the right path. Working with pens, a computer or a model is only a means to an end. The most important reward of constant practice is on an intellectual level. By leaving well-trodden paths, by trying out new ideas and designing by trial and error, you can tap into new veins of creativity and develop a diverse architectural repertoire. Developing creativity does not end with a university degree. It is a lifelong process that should be engaged in deliberately and intensely.

External influences not directly related to the assignment are also decisive factors in the design process. If the work on a design takes place in a team, ideas can emerge in dialogue with others, with each member of the team contributing to the process, advising others and finding the right path forward. The same holds true for an architect's interaction with a client, or with teachers evaluating assignments at the university. The exchange of ideas can help individuals grow beyond their limits, while the focused external feedback keeps them from throwing in the towel too soon and provides continued impetus. Students learn which methods do or do not help them achieve their objectives, and they benefit from the others' experience. › Chapter Creativity and creativity techniques

Spatial experience of architecture

Built architecture can also provide a wealth of experience. The intense study and physical experience of buildings is also a good way to become acquainted with ideas and methods. › Chapter Arriving at ideas While books introduce students to new worlds and serve as a source of inspiration during their studies, they are often quite selective and unable to present contexts in their totality. Students who take in a building with all their senses will have lasting memories and important experiences. It is essential to visit a building and experience it spatially, observing it from

Fig. 2:
The appearance of a building is affected by its subsequent users as well as by the initial design.

Fig. 3:
Local visits can show whether users have accepted the intended use of a building.

all sides in its surroundings. It is essential to touch and feel it, and see how people use it. › Fig. 2 This is the only way for them to get a comprehensive idea of the building and gain insight for their own work. Only if they experience a building themselves will it have a sustained effect on their work. › Fig. 3

Developing a perspective on design

The experience you have with design assignments and your reflections on built architecture will gradually help you develop a perspective on how to tackle a design assignment. The term "design perspective" refers to a conscious approach to designing and to the way you adapt designs to constructed reality. This need not involve an eccentric, idiosyncratic style in the sense of a distinct architectural language. Rather, a design

\\ Tip:
You should view as much built architecture as possible. A good way to start is by walking through your home city and carefully observing and analysing buildings on shopping or residential streets. This will give you a feel for the environment. It is equally important to study buildings designed by well-known architects. City tours, brief stops on trips, as well as excursions during or after your studies, all provide good opportunities to view the famous architecture in a region or city.

perspective is the unifying principle of a work and results from the way you deal with design assignments and projects.

This design perspective is often directly related to the designer's character, and is not limited to interaction with architecture. It can be an expression of an entirely personal worldview and associated with a broader social context or philosophy. Developing a design perspective is thus part of an individual maturation process and cannot be forced or artificially produced. When architectural students begin examining the architectural aesthetics and design perspectives of well-known architects, they are likely to look for role models and methods with which they can identify, and which they find adaptable to their work. They naturally find it helpful to understand the methods and perspectives already employed and to try them out or gain some experience with them in the designs they do at the university. This is the only way to explore recognized and famous worlds of design. But prospective architects should not shackle themselves to any single dogma that restricts their development and confines them to a certain path.

Fig.4:
Observing how a public square is frequented

Fig.5:
Light moods shift according to the time of day, and shade in a quiet area

DESIGNING IN CONTEXT

Each design emerges in a very specific context, whether it be a construction site and its surroundings, or a social and socio-cultural context.

While the process may begin with an examination of the site, the resulting building does not necessarily have to be adapted to surrounding conditions. An individual position or a counter-position can be formulated as an alternative. Even so, it is important to examine the site closely in order to understand the effects of certain decisions. Natural or anthropogenic influences will play a predominate role, depending on whether the site lies in a rural or urban environment.

Local presence

In most cases, the intense study of the site and its surroundings is extremely helpful in the search for a design idea. You should attempt to grasp the site three-dimensionally through sketches, measurements and visits, particularly if it has a distinct topography. You should also allow yourself enough time to study views of the surroundings and interaction with the landscape.

Landscape models

If a broader landscape needs to be considered or the topography plays a particularly significant role, it might be helpful to build a landscape model that shows elevations. This can be used to check and optimize the effects that the initial designs have on the surrounding space. It is also important to study possible views on visits to the site and to select an appropriate section of the model. › Fig. 6 If larger contexts are to be portrayed – urban systems, views between buildings etc. – all important relations should be incorporated into the model. › Fig. 7 When preparing designs in an urban environment, you should also conduct spatial analyses of the

\\Tip:
It can be advisable to visit the construction site and observe daily life at different times of the day. Where do pedestrians walk and from what perspective do they view the site? Where are quiet areas located, and where is there street noise? How are atmospheres created, and how does the light change throughout the day (see Figs 4 and 5)?

Fig. 6:
Landscape model: architecture in a rural context

Fig. 7:
Landscape model: architecture in an urban context

immediate and broader surroundings in order to get a feel for the location. These analyses can take the form of as-built plans, development structures, relations between streets and paths, the design of squares, green areas, and much more. › Fig. 8 In addition to the benefits mentioned above, the landscape model makes it possible to view the site from a distance on all sides and to discover connections that are often invisible from the site itself.

An examination of the location helps you understand the unwritten rules underlying the local situation. Systems, dependencies and relations between elements come into focus and cohere into a structure that can serve as a foundation for a design. The design can be harmoniously integrated into this structure or employ alternative approaches to interpreting it. Likewise, you can deliberately choose a "confrontation" with the surroundings or develop an autonomous design idea. It is essential that the work be based on an organic understanding of the place. If you ultimately seek a confrontation with the surroundings, this will be a conscious choice and should be comprehensible as such.

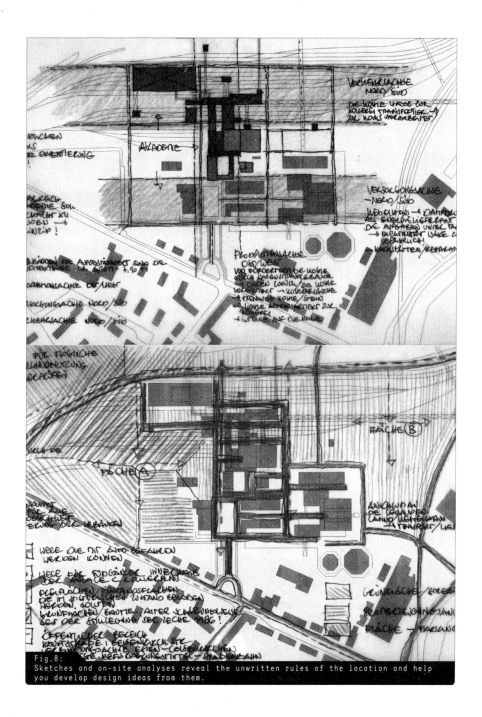

Fig.8:
Sketches and on-site analyses reveal the unwritten rules of the location and help you develop design ideas from them.

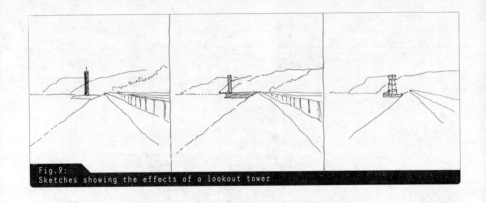

Fig.9:
Sketches showing the effects of a lookout tower

LANDSCAPE AND CLIMATE

A study of the broader landscape extending beyond the individual site can yield a variety of strategies, from the design of a distinct landmark to efforts to simplify the visible, constructed building and adapt it to the surroundings.

Topography

Site typography determines how the design will be integrated into the landscape. Regardless of whether the site is absolutely flat, sloped, terraced, tiered or hilly, topography will always affect the building and the subsequent interrelations between interior and exterior space. The type of terrain can also influence the layout of floors inside the building. › Fig. 10 For instance, height differences in the landscape can be continued inside the building, and entranceways can be positioned to use access routes to public street space. › Chapter Designing in context, Urban planning and architectural context Naturally, if parts of the design make it necessary, the site can also be landscaped. In general, a building can be designed to respond to the topography or even be playfully adapted to it. However, the architect can also make a conscious decision not to adopt such an approach and thereby create a self-contained unit that functions independently of the topography and establishes a clear separation between the surroundings and the architectural intervention.

Sloping sites

When there are differences in elevation on a site of more than one storey, you must consider how the internal structure of the building is to respond to the topographical situation and what relationship should be established between the interior space and the surrounding area.

A building on a sloping plot can be set into the hill, project over the slope, respond to the slope with tiers, or modify the slope. › Fig. 11 These

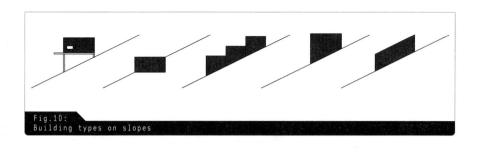

Fig.10:
Building types on slopes

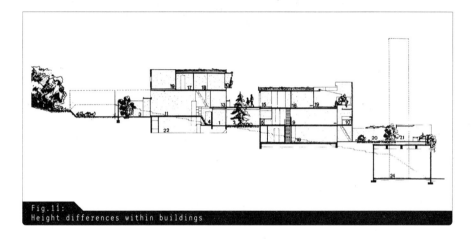

Fig.11:
Height differences within buildings

different forms give rise to diverse relations between interior and exterior space. The special features of the topography, particularly on complex sites, often inspire intriguing design ideas. > Fig. 12 If there are broad views between the site and the surrounding landscape, you should determine both the angle from which the building will later be seen and the possible connections that might prove interesting in the landscape. > Chapter Designing in context, Urban planning and architectural context

Climate-responsive architecture

Aside from site topography, an analysis of climatic conditions provides a further method for developing a design. Depending on the regional climate, it may be advisable to use the sun to determine the orientation and organization of a building. The building can be open on its sunny side, allowing solar energy to enter and be stored inside, or it can be closed off from the sun to keep heat out of the building.

Fig.12:
The view of the landscape as a basis for a design

Furthermore, the construction method, the materials and the architectural form can be adapted to the macro- and micro-climatic conditions of the location. In warmer regions, buildings can be constructed as heavy structures or be set into the ground in order to use the storage capacity of the materials or the earth for cooling purposes. Alternatively, architects can facilitate cross-ventilation by exploiting the prevailing wind direction. In contrast, in moderate climate zones, the building's area/volume ratio can be optimized to minimize the loss of heat through transmission and to maximize the amount of sunlight falling on the façades.

URBAN PLANNING AND ARCHITECTURAL CONTEXT

In an anthropogenic environment – one constructed by human beings – human factors usually exert a greater influence on a design than natural ones. Anthropogenic factors can be analysed in a detailed study of the environment with the goal of developing possible design approaches. The surrounding area usually includes neighbouring buildings, streets or trees that form points of reference. There might also be additional buildings on the property that you will need to integrate into the design. If buildings abut the site, you will also need to clarify whether your building may (or must) be built directly alongside them. In densely developed areas where buildings stand on the edge of streets, you must also consider whether you want to adopt the form of the adjacent buildings. If the building fills a gap in a row, it might be necessary to respond to and interpret the different heights of the adjacent buildings. › Fig. 13 If the building

Fig.13:
Building roofs influence the appearance of the urban environment

stands as a solitary structure on the street, you can adapt the various existing systems or provide a counterpoint to them. Typical parameters in this type of analysis are:

_ Roof type
_ Building orientation
_ Distance between the street and the buildings
_ Materials
_ Window type and size

Interaction with existing buildings

Most architectural assignments in the future will no longer centre on creating new buildings on undeveloped tracts of land. Even now, there is a growing demand for ideas and plans of how to use existing buildings. Such assignments give you the chance to derive strategies that reflect the architectural distinction between the new and the old. You can put existing structures to new use, adapt them to suit modern needs, transform buildings, and give them entirely new identities. You can keep or add to developments (landmarked structures, flexibly used buildings etc.) or even individual building elements (façades, structure etc.). › Fig. 14 Alternatively, if remodelling and assigning new functions to existing buildings proves too costly, they can be completely demolished. In this process, an important question is the purpose of the architectural interventions: should old and new exist independently of each other as two distinct parts, or should they enter into an architectural dialogue? Should architects strive for extensive uniformity in a development or does quality a product of emphasizing differences? › Fig. 15 It is always crucial to ask whether demolition can be justified, or is even necessary, from an economic, environmental and cultural point of view. › Chapter Designing in context, Social and socio-cultural factors

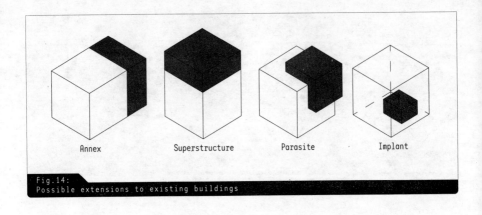

Fig.14:
Possible extensions to existing buildings

Fig.15:
A sketch analysing the use of the existing development structure

Urban development parameters

To avoid chaotic growth and the destruction of historically evolved structures, most countries issue detailed regulations on property development. These include:

_ Restrictions or specifications concerning the building area or floor space
_ Regulations on frontage lines and the boundaries within which a building can, or must, be built
_ Number of floors
_ Specifications on roof type and construction methods
_ Regulations on entrances and circulation
_ Distances to other buildings and property boundaries
_ Protection of trees etc.

If you are designing a building that will in fact be constructed, these regulations must usually be observed. Furthermore, urban planning conditions need to be studied at an early stage of the project to ensure that the resulting design can indeed be built. Nonetheless, if regulations are strict and allow little freedom, there is a risk that familiarity with them will impede creativity in the design process. It is important to find a middle ground between the free generation of ideas and their implementability. This will determine the details you need to study and the depth to which you must go in your analyses.

Urban references and axes

Densely developed urban environments often contain reference points and structures that, as overarching urban principles, have determined the design of most of the surrounding area. › Fig. 16, 17 As part of an urban planning analysis, you should study the role and significance that your building's use or function will have in the broader urban structure and the immediate surroundings. You can also derive basic urban "figures" from this study as a point of departure for additional analyses.

For instance, if adjacent buildings are set back from the edge of the street, you can design a projecting building to create an urban accent. Alternatively, by designing buildings that are themselves recessed from the street, you can create a forecourt or a courtyard space. If the surrounding area already contains a number of dominant solutions whose axes and forms are incorporated into your design, the building will be well

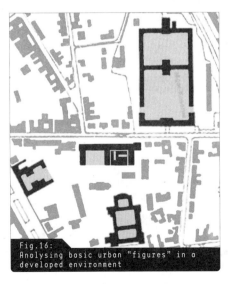

Fig.16:
Analysing basic urban "figures" in a developed environment

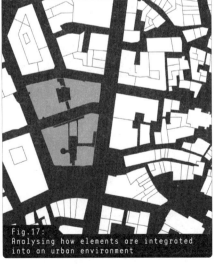

Fig.17:
Analysing how elements are integrated into an urban environment

23

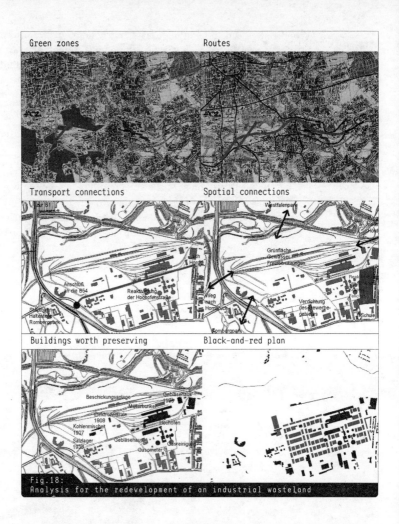

Fig. 18:
Analysis for the redevelopment of an industrial wasteland

\\Tip:
In addition to studying local conditions, you can use area, city or land registry maps to conduct a variety of analyses (see Fig. 18). An as-built plan that shows the surrounding forms as blackened boxes will draw your attention to the urban building network and open spaces, while street maps will reveal important relations (see Fig. 19).

\\Hint:
For more information on the design and typology of public squares and spaces, we can recommend *City Planning According to Artistic Principles* by Camillo Sitte, Columbia University Press.

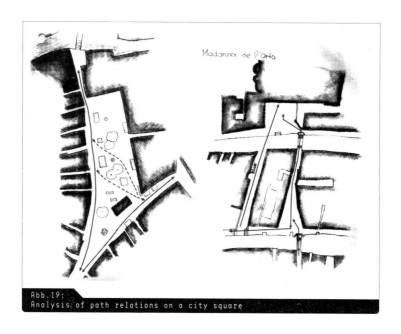

Abb.19:
Analysis of path relations on a city square

> ✎ integrated into the surroundings and represent a subdued urban planning solution.

Depending on the surrounding area, important reference points or freestanding structures can provide a basis for dimensional relations and axes: an asymmetrical square creates edges with various angles, while an orthogonal street layout opposite the site opens up a view of the building. Alternatively, you can create a counterpart to an important element in urban space. There are diverse ways to communicate with urban sur-
> 🏛 roundings through design.

Accessing the property and the building

Normally, you will already know where and how people access a property. The site either lies on a street, or has a path or road leading to it. Independent of functional requirements, access routes also determine they way users, visitors and passers-by perceive the building. > Fig. 20 An important question to consider is what effect the design will – or is supposed to – have on people approaching the building. > Chapter Designing in context, Social and socio-cultural factors

One important point is the height of the building's entrance level in relation to the street. If the entrance level is below street level, the pathway

Fig. 20:
Access by means of a bridge, or via an axial staircase and portico

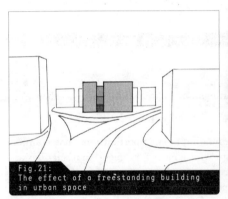

Fig. 21:
The effect of a freestanding building in urban space

Fig. 22:
Access across a residential courtyard blending public and private spaces

leading to it will seem to be of secondary importance. If the level is much higher, it is more likely to be perceived as "grand" and to inspire awe. If, on the other hand, a building is entered via a front square, the entranceway will create a sense of distance, even though it may appear more impressive. An entranceway through a courtyard divert attention from public space and create a commons area in front of the building. › Figs 21, 22 You should also consider placing the entrance to a building at ground level for wheelchair users, particularly for public buildings.

Orientation

Building orientation is another point to consider along with the building's three-dimensional shape, its position on the site and its entranceway. A building can have a sealed-off and solid effect, or seem trans-

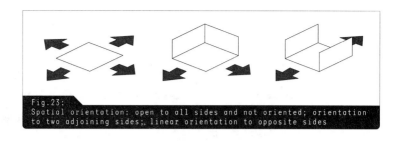

Fig.23:
Spatial orientation: open to all sides and not oriented; orientation to two adjoining sides;. linear orientation to opposite sides

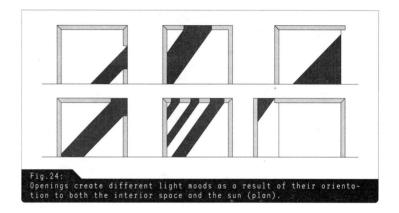

Fig.24:
Openings create different light moods as a result of their orientation to both the interior space and the sun (plan).

parent and open on all sides. › Fig. 23 But these different possibilities lack orientation, that is, they will convey the same impression to all sides.

Depending on the site, you may want to treat the building's individual sides differently. You may, for example, desire to orient the building toward the sun, or need to design rooms with different lighting requirements. › Fig. 24 and Chapter Designing in context, Landscape and climate In moderate climate zones, orienting living space toward the sun is regarded as desirable, but it might prove disadvantageous in an artists' studio or a museum, since these spaces require uniform, diffuse northern light.

Other factors influencing a building's orientation can be found in the architectural environment. If the street side of the building is loud and lively, you may need to shield the residential and leisure spaces. The rear of the building may have an open park view that users would like to experience from inside the building. Or perhaps a large number of apartments are meant to benefit from a special feature of the landscape, such as a

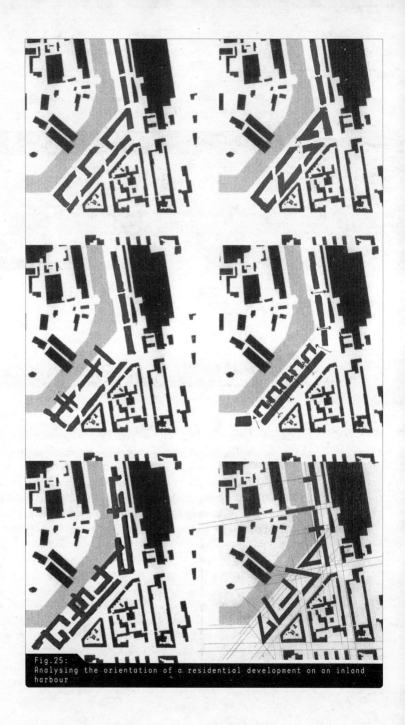

Fig.25:
Analysing the orientation of a residential development on an inland harbour

Fig. 26:
Orchestrated views of exterior space

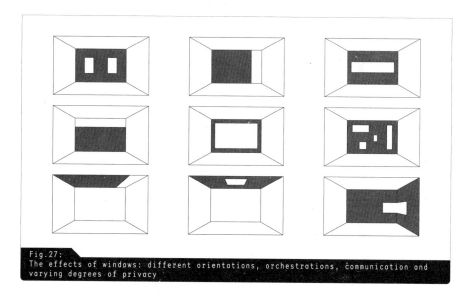

Fig. 27:
The effects of windows: different orientations, orchestrations, communication and varying degrees of privacy

nearby waterway. › Fig. 25 These different situations may require a specific building orientation and inspire location-related approaches to developing the design. Examples include an entire wall of glass to emphasize a special view in an otherwise closed-off room, or perhaps a series of small windows that show specific details of the surroundings as an integrated composition. › Figs 26, 27

The relationship between interior and exterior space can lead to the development of a variety of architectural principles, from "closed/introverted"

Fig.28:
Integrating the exterior space into a church space and its use

Fig.29:
Glass surfaces can emphasize interaction (through transparency) or convey a sealed-off quality (through reflection).

to "open/interactive." A glass house in a dense urban environment is probably ill advised due to its lack of privacy, but it may prove a good idea in a vast empty landscape where the building – as a minimal architectural intervention – will enter into a dialogue with the surrounding area.

Interaction between buildings and their surroundings

If the building is meant to relate to its surroundings, the way it does so can be based on the external perception of the building or on the interaction between interior and exterior space. For instance, a panoramic window can be used to emphasize a lovely view of a valley, or a carefully positioned window can transform an interesting visual axis into an effective element in the building's interior. A building can also be built around a prominent tree, which will then become part of the interior space. › Chapter Designing in context, Landscape and climate

The use of wall-to-ceiling windows or specific views into and out of the building in order to differentiate open and closed spaces is an important design tool for creating a range of spatial experiences. › Figs 28, 29 and Chapter Design and function, Responding to user needs

Differing elevations and views across the inside of a building can also play an important role in creating concepts. If outside areas are not shielded from views, the ensuing lack of privacy quickly diminishes their appeal. However, an interesting view from a shielded area can dramatically alter spatial impressions. The same is true of designs that optically conceal the boundary between interior and exterior space as a means of integrating the private exterior space into the building. › Figs 30, 31

For this reason, a building should never be considered only in isolation. If possible, you should examine the interaction between the property, the building and its use. A design is influenced by many location-related elements, such as shadows from adjacent buildings, sunlight at different times of the day (and during different seasons), as well as the provision or hindrance of specific views into and out of the building. These different aspects also show that a design always emerges from the interplay between the location and the diverse architectural requirements.

Fig.30:
Interior space with a fluid transition to exterior space

Fig.31:
Shielded exterior space with a view

SOCIAL AND SOCIO-CULTURAL FACTORS

The previous chapters addressed many factors that influence how people perceive architecture. In this connection, it is important to distinguish between the way architecture is perceived by the senses and the way these impressions are processed in the human brain. Even if sensory perception always remains the same, information is processed in different ways due to the differing individual thought patterns, which are, in turn, strongly influenced by personal experiences and social and cultural contexts. This explains why people will perceive an event or a building in different ways. Generally speaking, perception is not objectifiable, even if an "objectified subjectivity" emerges from the combined opinions of the great majority. This subjectivity is mostly confined to a narrow social, historical and cultural framework, since people are largely socialized and influenced by their education. An examination of the historical, social and cultural backgrounds of an assignment often yields important insights and can provide a source of inspiration for generating ideas.

Historical factors

The study of a location should not be limited to an examination of the immediate environment. Every architectural project is a response to the history of a place and plans its future. Designing and modifying an existing situation is an intervention that is inevitably perceived by those in the environment as a part of a continual process. Buildings are constructed and used for specific purposes. They can then be used in new ways, remodelled, added to, torn down, rebuilt or recycled. At times they may remain vacant or fall into disrepair. › Fig. 32 Their use often has a social significance, regardless of whether it is embedded in a broader history that is later recorded in books or in the more personal stories that people associate with specific places. Efforts to create references to the history of a place can lead to various approaches to developing design ideas. You can refer to collective memories – impressions made on the memory of an entire society – or to the entirely personal stories of the client or the previous users of the property. › Fig. 33 Similarly, you can gain personal access to a subject matter by incorporating your own experiences into the initial idea.

It is important to ensure that the reference system reflects the significance of the architectural assignment. For example, if the assignment has social implications – such as a design for a museum or a memorial site – it may be advisable and even necessary to focus on the link to historical events. But when the object of the design is a residential building or a shopping centre, you should give careful thought to whether a historical reference is appropriate to the context.

Fig.32:
The addition of a glass hall creates a new use for an old castle ruin

Fig.33:
Urban dialogue between a new cultural institution and a Roman temple

Fig.34:
Residents using public and common areas

The socio-cultural context.

If your design idea is a response to social developments, you can derive basic approaches from either an overarching social context or specific phenomena. For instance, when conceptualizing a residential complex, you can attempt to counteract the increased privatization of public space by creating public pathways leading across the site or by integrating public facilities. › Figs 34 Further, barrier-free accessibility to all parts of a building can have an impact on both urban development and access to the building.

One interesting aspect of socio-cultural analysis is that it gives you the chance to express your own views and philosophy: what mix of uses is necessary to integrate elderly people or those of different ethnic backgrounds into society? How much community and how much individuality are possible, and necessary, in our shared lives? What role does environmental and climate protection play, and how can its importance be reflected in the design concept? What limits are there on development density if we wish to avoid mutually adverse effects and prevent the uncontrolled development of the landscape from worsening?

Whatever approach you choose, it is crucial that you relate the initial architectural measures to a broader context and value system. This will help you develop your own perspective.

Regional architecture

Over the course of time, diverse regional architectural forms and typologies have evolved from different climatic, cultural and social conditions, as well as from the often limited availability of materials and from specific uses or utilitarian forms. ＞ Fig. 35 and Chapter Designing in context, Landscape and climate

Often, these architectural typologies offer astonishingly simple solutions to a variety of needs. Examining and analysing regional architecture can frequently lead to surprising new insights on the complex interrelations between architecture and architectural and social contexts. It is important not to view architecture in isolation from its function, its surroundings, and the period in which it was built.

Internationalization

In general, we cannot compare modern conditions to those under which many traditional architectural forms evolved. We have many more possibilities and materials for coping with challenges, even those involving difficult climatic conditions. Uses and requirements have also changed dramatically. As a result of the ongoing internationalization processes, the differences between places and cities seem less pronounced today than in the past. Materials and architectural forms have not only become increas-

Fig.35:
Regional architecture in Mediterranean cultures

ingly similar, but our habits and behaviour have also become more and more alike due to our knowledge of other countries and cultures.

To develop sustainable architecture that enters into a lively dialogue with its surroundings, you should become acquainted with and respond to differences in regional and local architecture. Traditional architectural forms and typologies play a major role in this process since they strongly influence the way a new building is seen in its environment. For example, because of local architectural traditions, a new brick structure erected in the north of Germany or in the Netherlands will fit in seamlessly with the surroundings and have the appearance of a traditional building. Yet the same structure will stick out like a sore thumb in a town in the south of Italy where the buildings have monochrome rendered façades. The different moods of light in the Mediterranean region – marked by rich contrasts and warmth – make simple structures with three-dimensionally shaped façades appear much more dynamic than similar buildings in, say, Moscow, where there is a low-contrast light environment. Modern architectural forms are not randomly interchangeable or reproducible, and they are more closely intertwined with a place than they may seem at first glance.

Symbolism and iconography

Architecture imparts information. This information can be perfectly obvious and recognizable to everyone, or it can hidden from view and only gradually reveal itself to the observer. In all periods of architectural history, people have discussed how architecture needs to convey information and how this information is related to the purpose and use of a building.

Two opposing positions can be made out: one school strives for a unity of content and form, following the principle of structural honesty. The other emphasizes the role played by architecture in carrying meaning and creating identity, independent of its purpose and functionality. This school uses the symbolism and iconography of individual architectural motifs to do justice to this idea.

In eclecticism and postmodernism, architectural references and motifs are used to trigger associations in viewers and achieve specific effects. For example, a building may appear more impressive if it incorporates the formal vocabulary of the Greek temple or the icons of modern architecture. The design will then allow viewers to draw conclusions about the building's purpose, function and owner. It is unimportant whether these assumptions are true or not. Outer shell and function are not supposed to form a creative whole. Rather, the architects who design in this way strive

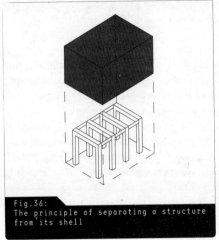
Fig.36:
The principle of separating a structure from its shell

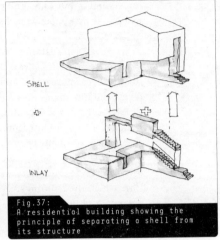
Fig.37:
A residential building showing the principle of separating a shell from its structure

to separate the outer shell and give it a measure of independence so as to emphasize its role as a bearer of meaning. › Figs 36, 37

At the same time, architects can also make symbolic references to thematic parallels over and above formal references. For instance, during the Renaissance, there was a return to the values and view of mankind that prevailed during antiquity. In architecture and art, this was expressed in new interpretations of the buildings, principles and themes of antiquity. Nonetheless, it must be noted that a message that appears logical from a personal or professional perspective may be interpreted in an entirely different way by others. This is why you must analyse the cultural and social context before using stylistic devices such as symbolism and iconography.

› Chapter Constituents of design

\\Hint:
Symbols carry meaning, and their value lies in the way they unify content, meaning and form. Symbols represent an object while preserving its outer and inner unity. By contrast, when architects "cite" certain motifs, they do not aim for a unity of content and form.

\\Tip:
Architectural symbolism is an extremely exciting and complex topic. For a more in-depth study, we recommend *Learning from Las Vegas* by Robert Venturi and Denise Scott Brown, MIT Press, 1977.

Links to other disciplines

The relationship and interaction between architecture and other artistic disciplines can best be understood by examining the various periods of art and architectural history, remembering that art and architecture do not always apply the same principles and methods in any given period. One of the most important differences is that, aside from its artistic ambitions, architecture must satisfy the elementary need that all people have for a place to live, work and sleep, and for protection from the weather. Even if these special functions give architecture a special status in relation to the other arts, there exist connections with these other disciplines that can serve as a basis for producing new design ideas:

The periods in which art and architecture were closely related show most clearly which principles can be transferred from one field to the next. In the Renaissance there were significant parallels between the articulation of foreshortened space in paintings and the selection of perspectives in constructed reality. The goal of the Bauhaus was to combine architecture – as a synthesis of the arts – with all the other artistic disciplines to unify all the arts. The formative themes of postmodernism, including iconography and identification, are most clearly legible in its architecture. Numerous parallels also exist in contemporary art.

In his accessible spatial sculptures made of corroded steel plates, Richard Serra explores the nature of the material by showing its aging process. He also emphasizes its heaviness, with many of the sculptures weighing upward of 100 tonnes. > Fig. 38 The powerful spatial and atmos-

---

\\Hint:
The arts are generally divided into four different categories:

The visual arts, which include painting, sculpture, architecture and the applied arts
The performing arts, whose main disciplines are acting (theatre, film) and dance
Music, broadly divided into vocal and instrumental music
Literature, which is subdivided into narrative prose, drama and poetry

Works of visual art are usually spatial, physical objects that have an effect in and of themselves and do not require an interpreter.

\\Hint:
In their spatial and physical effects, outdoor sculptures display many parallels to architecture. The interesting interactions between an object and its surroundings can be used to train spatial perceptions. The parallels are particularly striking if the artworks have been created for a specific place.

Fig.38:
Sculptures by Richard Serra

pheric effects of these sculptures are reflected in the unease that they initially caused in public space. Peter Eisenman, among others, has taken a comparable approach in architecture, particularly in his design for the Holocaust Memorial in Berlin.

Even so, most works of art cannot be understood through observation alone. It is also important to read texts by artists, as well as secondary literature and biographies. Most artists have written extensively about their working methods and their motivation. It can be very rewarding to experience and try out work and design methods from the other artistic disciplines. By examining the various techniques of painting, sculpture, photography, music and other art forms, you can find inspiration for your own architectural work.

DESIGN AND FUNCTION

The function of a building will usually have a formative effect on both its design and the way the design is developed. Depending on the your approach, function can provide a general framework that you must adhere to, or form a point of departure for design ideas. Many architects develop their buildings by designing a floor plan or spatial framework to fulfil certain requirements and functions. Drawing on their creative skills and experience, they derive from this a specific expressive style and formal vocabulary that transcends mere functionality and aims for a unity of form and function.

Function as the point of departure for a design

Ever since the modern period, it has been common to use function as the basis for designs, and function has also assumed special importance in architectural studies. Creating a design that reflects and visually expresses function is a fundamental architectural objective.

The steps described in the sections below are offered as tools and possible approaches to help students understand function and to incorporate it into their designs. In principle, these steps are not required to create a "functioning" design, yet it is advisable to use such tools, particularly when the function in question is not one with which you are acquainted in detail. The term "tool" already emphasizes that these aids are not at the centre of design work. Illustrating mutual dependencies in a functional diagram presupposes that you already fully grasp their intricacies. One way to understand them is to read publications that provide information on typical spatial needs or functional connections. ˃ Appendix, Literature In addition, you can analyse existing buildings with identical or similar uses. Studying the plans of several exemplary buildings will show similarities and common rules as regards the layout, size and structure of the individual areas. These can be incorporated into your design.

RESPONDING TO USER NEEDS

When turning your attention to a building's function, you usually also have to deal with the people who need it, whether in the form of housing, workspace or recreation. It is important to analyse how users practise this function within the building, or how they wish to experience it. If we consider a cinema, for example, we can see that people expect a specific world of experience from this structure. They want to be entertained and perhaps even immersed in another world for a short time so that they can forget their everyday lives. Your design can take these expectations into account. By contrast, you should design an office workplace in such a way that employees can concentrate and are not interrupted during their work. You need to meet their expectations about lighting, ventilation, acoustics and workplace design.

There are different ways to study user needs: on the one hand, you may be personally acquainted with the user and thus know the person for whom you are designing the building. On the other, the design assignment may be focused on a specific target group (e.g. senior citizens who live in planned apartments) as opposed to an actual person.

Individual needs

If the user is known, you can study his or her individual interests and needs. For instance, if the assignment is to build a home and a studio for an artist, discussions with the artist and observations of his or her working methods can yield insights into requirements and needs. Perhaps the artist wants a quiet, uniformly lit and shielded studio space, or enjoys the view of the landscape or the hustle and bustle of the big city.

Focusing on a target group

You will face an entirely different situation when designing a structure for a specific target group. In this case, you must first get an idea of the group's general requirements. Here, it may be helpful to study local solutions and buildings in order to learn more about the target group's needs, or to analyse existing problems that exemplary projects with the same function have confronted. These problems can then be addressed in the design. › **Chapter Arriving at ideas**

Contact with users

In most cases, you can examine user needs by contacting them directly and sharing ideas with many different types of users. You should resist the temptation of attempting to understand internal processes (e.g. at specific production sites or a fire station) from the "outside" only. This is particularly important with functions of which you have little knowledge. The people who deal directly with a building's function on a daily basis will have a different view of things, and it is important to consider their views when you work together to adapt functional processes to spatial designs. The intense exchange of ideas with employees or users on all levels of an organization will help you analyse processes, structures and, most of all, problems arising in existing buildings. Even so, information obtained solely from users is of limited use in a design because they lack an architects' background knowledge of design. This is why working together on utilization concepts and detailed planning is the ideal way to design a building that functions well down to the last detail.

SPATIAL ALLOCATION PLAN AND INTERNAL ORGANIZATION

Spatial requirements are usually given for definite construction projects. For instance, a family will have a specific idea of the house they want to have designed, a company will need workspaces for a given number of employees, or a museum will require exhibition space for specific exhibits. The necessary spaces and volumes can be derived from these needs and can provide an initial idea of a project's size and dimensions.

Space and volume

For standardized uses such as apartments and offices, you can arrive at a rough estimate of building volume by adding the space of the building's structure and secondary rooms, expressed in percent, to the required floor space and multiplying the result by a standard ceiling height.

Structures such as indoor swimming pools, museums and event halls usually include a variety of rooms, some of which have entirely different requirements for space and room height. This is why it is important to look at each area separately in order to address specific needs, and to design special features in the creative process. An indoor swimming pool might be a dominant structure in an ensemble of buildings, so that its subordinate functions can either be included in that structure or in other separate facilities.

If you decide to design in this manner, it is important you do not view the rough attempts at creating a structure as the actual design. Otherwise you run the risk of accepting the rough form as a given and not exploring the design in any greater depth. You should always keep in mind that the structure is an abstract figure and its form can be further modified as desired. › Fig. 39

In addition to making rough estimates of space and volume, it is important to create a spatial allocation plan that not only specifies the

\\Example:
Say a family needs about 120 m² of living space. To compute the total amount of space, add 20 to 25 percent for the space taken up by walls and shafts. Multiplying the total space by a floor height of 3 m yields a volume of approximately 450 m³. If you are designing a two-storey cubic building, this translates into an 8 x 8 x 8 m cube. Mental exercises like these can give you a sense of building size.

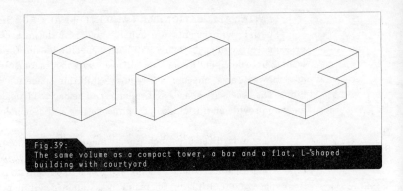

Fig.39:
The same volume as a compact tower, a bar and a flat, L-shaped building with courtyard

Tab.1:
Example of a spatial allocation plan

Consulting centre

Function	Room size [m^2]	Number	Total space [m^2]
Reception area/assistants	40	1	40
Waiting room	40	1	40
Assistants	12	2	24
Director	40	1	40
Office	20	1	20
Group manager	25	2	50
Consulting room	25	17	425
Conference	30	1	30
Staff room	20	1	20
Toilets	40	1	40
Secondary sales area		15%	109
Total		**28**	**838**

individual functions or rooms and their spatial requirements, but also divides rooms into thematically related groups. › Tab. 1

Spatial allocation plan

For many design assignments, the spatial allocation plan is already provided. Even so, the design assignment may sometimes involve developing the appropriate function together with the spatial allocation plan. Practising architects, in particular, often encounter clients who have a clear idea of the primary function of the building, but do not know what additional spaces and secondary functions are needed.

Therefore, your first job is to determine the main areas of use as well as the required secondary areas such as the lobby, toilets, entrances and

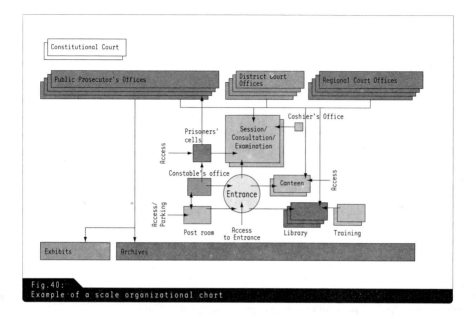

Fig.40:
Example of a scale organizational chart

hallways. You also need to figure out the spatial needs of each area if you want to gain an overview of the entire design assignment.

If, for instance, a company wants to create office space for 500 employees, you must first determine the office type (individual offices, open-plan offices) required for the employees as well as the amount of space each one needs. The secondary functions and access routes must also be taken into account to get an overall impression of spatial needs.

You can take a similar approach to, say, designing housing for elderly people. Together with the client, you must initially determine whether individual housing functions need to be integrated into the residential units, or whether they should be organized centrally for all the occupants.

\\ Hint:
For more information on housing functions, please see *Basics Design and Living* by Jan Krebs, Birkhäuser Publishers, 2006.

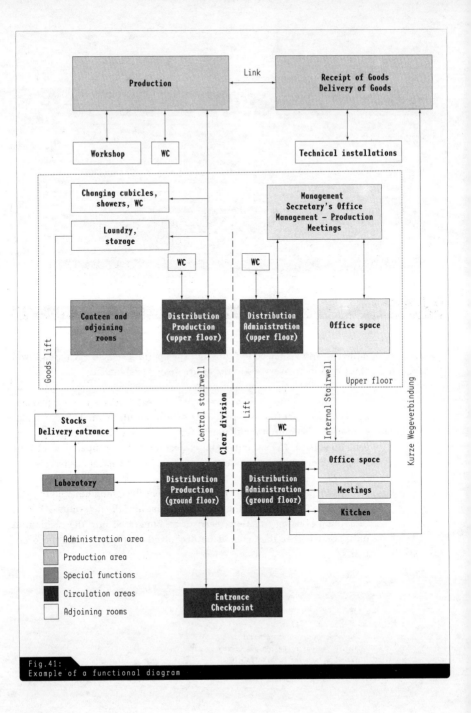

Fig.41:
Example of a functional diagram

The spatial allocation plan makes it possible to form groups of functions or spaces that have the same requirements as regards room height, lighting etc. It indicates the proportionate arrangement of spaces and will give you a feel for the size and proportions of the design.

Interior circulation and lighting

To assess the design further, you can develop circulation and path systems that structure the building and provide information on its volume and shape. For example, if an office building consists of individual offices, you can examine how far natural light falls into the building as a way of determining the building's depth. This can then serve as the basis for designing the building structure.

Internal organization

It can also be helpful to give some thought to the functional organization of the assignment. This should be expressed visually in order to determine the functional links between the building's various rooms and areas. One common tool is a functional diagram that graphically depicts all the functional areas or rooms and shows their interrelations. › Fig. 41

The diagram visually represents the building's internal organization. It reveals areas and connections that may have a formative effect on the design. For instance, if a private or sensitive functional area has to be separated from public or semi-public spaces, the resulting spatial layout may strongly influence the design.

Architectural assignment and purpose

It is important that you not equate the predefined spatial allocation plan with the actual architectural assignment and the purpose of

\\ Tip:
At times it may be helpful to draw the rooms in the spatial allocation plan as true-to-scale spaces in different proportions. This will give you a visual sense of the arrangement of spaces in the design assignment.

\\ Example:
Say an office building is a double-loaded structure with a middle corridor and a ceiling height of about 3 m. The maximum depth of the building can be determined by examining the maximum depth that an office can have and still be illuminated by natural light (approx. 5.5 m) falling through floor-to-ceiling windows. If you add the width of the central corridor (about 1.80 m), the thickness of the exterior walls (about 40 cm) and the corridor partition walls (each 15 cm), you will have a building depth of about 14 m. If the offices are arranged as combi-office zones, you can create a building depth of up to 16 m.

45

the design. In fact, you should critically examine the plan and modify it if necessary. An architectural assignment transcends the pure function of a building. Moreover, the approaches described above should always be seen in relation to the other determinative factors.

For example, a community centre will require not only a large hall with the affiliated secondary rooms and the necessary infrastructure. In order to fulfil its purpose, it must also be a place of encounter for diverse people and groups. So to design it successfully, you must grapple with soft factors in addition to the way processes and spaces function. These soft factors include accessibility, openness, lighting, atmosphere and views. These factors often cause conflict between the architectural assignment and the spatial allocation plan, which can be overcome by adapting the plan.

CONSTITUENTS OF DESIGN

When working with forms and design elements, you should always bear in mind that it is impossible to predict with any certainty the impact a design will have on others. A design is good not only because it satisfies the entire spectrum of individual requirements, but also – and more importantly – because the structure created defines the relationship of the individual elements to one another and thus arranges them in a new order. Exploring every possible solution by making sketches, drawings and models is the only way you can continually review your ideas and solutions.

The following chapter focuses on design as the initial point of departure in the design process. › Chapter Design basics

ORDER AND PROPORTION

Since Antiquity, proportions have been used to design façades and entire buildings. Their application can be seen in buildings throughout the ages: from the construction of temples in the ancient world to the building of medieval churches and Renaissance villas, and from the architecture of classical Modernism to present-day structures.

Architects have repeatedly attempted to express ideal proportions with the aid of mathematical formulae. During classical Antiquity, they analysed the proportions – especially in relation to dimensional ratios – of their temples and building elements (e.g. the different column orders). They then developed them further. Many Greek master builders and architects had a profound knowledge of geometry and used numerical ratios to arrange individual building elements in relation to one another when building temples.

\\Tip:
The historical development of structural proportions is a fascinating subject. As you study this field, you will discover interesting parallels and developments, which have lost none of their relevance today. It is worth reading the books by Vitruvius, Alberti and Palladio, if you want to study the historical development of the laws of proportion in architecture (see Appendix, Literature).

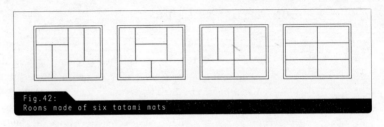

Fig. 42:
Rooms made of six tatami mats

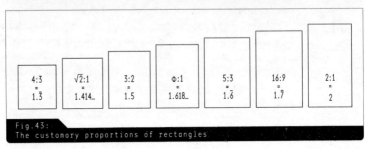

Fig. 43:
The customary proportions of rectangles

The double meaning of the concept of proportion is already contained in Vitruvius' definition (30 BC). On the one hand, the term "proportion" refers to the relationship of the parts to one another, and on the other, to human beings. For instance, the dimensions of the lower diameter of a Doric column in relation to its height form a ratio of 1:6, thus corresponding to the proportions of the male human body. The ratio of the height of an Ionic column to its diameter is 1:8, thus reflecting the proportions of a woman's body.

Japanese tatami mats

Japanese tatami mats provide a simple yet effective example of the application of proportional principles in architecture. Tatami mats are the basis upon which traditional Japanese houses are built. These mats generally measure 85 cm × 170 cm, although their dimensions may vary from region to region. Because of their 1:2 ratio, the mats can be arranged in an endless variety of constellations. The mats also determine both the size and proportions of rooms. (The standard Japanese room generally consists of six tatami mats.) › Fig. 42

In addition to the proportions of tatami mats, there are other proportions that are used in architecture and many other areas. › Fig. 43

The golden section

One of the best-known proportions is the golden section, which expresses the relationship of two numbers or lengths in a ratio of approximately 1:1.618. The golden section is an irrational number, like the

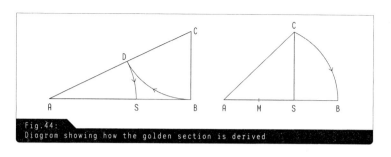

Fig. 44:
Diagram showing how the golden section is derived

mathematical symbol π, because it cannot be expressed as a fraction of two whole numbers. Mathematically, the golden section is defined by the formula:

a/b = (a+b)/a

It can also be constructed with the aid of a circle. › Fig. 44

The Modulor

The introduction of the metric system ended the dominance of a wide variety of units of measure. Unlike the old units of length (inch, foot, yardstick and rod), the metre is related to the earth's circumference and not to human proportions. As a result, one of the two meanings of proportion – the direct relationship between the "modulus" and the human being – ceases to apply. Proceeding from a desire to reintroduce human scale into architecture, Le Corbusier developed his own system of measurement (the Modulor) from the proportions of the average human being and the golden section. The Modulor makes it possible to design proportions and dimensions that are directly related to their use, for example, in table and balustrade heights, as well as for the proportions of windows, entire rooms and façades. › Fig. 45

Experiencing proportions

Aside from all the mathematical approaches and analyses, proportions create a subjective feeling of wellbeing in the viewer. Even though everyone can understand the desire for both systematization and directly applicable rules, a great many seemingly well-designed buildings and

> \\ Hint:
> If you want to learn more about the development of the Modulor and the background to the modular measuring system, see Le Corbusier: *The Modulor*, vols 1 and 2, Birkhäuser Verlag, Basel 2000.

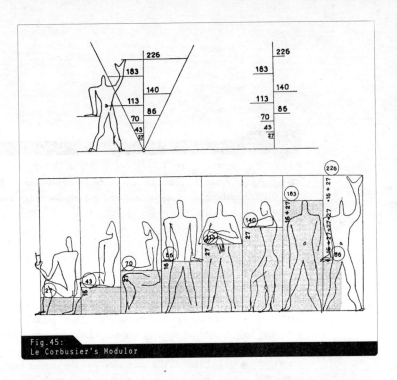

Fig.45:
Le Corbusier's Modulor

façades are not based on any comprehensible mathematical rules. Thus, achieving balanced proportions always also means that designers must feel their way forwards when they determine the dimensions and compositions of various building elements and the relationships between them.

For example, if you are working on the proportions of a façade, you will discover that there are a number of familiar rules and proportions you can fall back on. Even so, in many cases, façades only have well-designed proportions once you start to interpret and apply rules and proportions, or experiment until you arrive at the desired results. This may involve shifting and altering, say, a large panoramic window until you are satisfied with the proportions of the interior and the façade and they finally harmonize with the overall design.

BASIC ELEMENTS OF DESIGN

Geometrical forms

The geometry taught at school deals with points, straight lines, planes, distances, angles etc. Working with the axioms established by Euclid (approx. 365–300 BC) and subsequent refinements of his theories, you can derive forms that will serve as a basis for your design. The two-dimensional forms (planes), for example, include the triangle, square, rectangle, circle and rhombus; three-dimensional forms (bodies), on the other hand, include forms such as the cube, rectangular parallelepiped, sphere and cone. You can use these basic mathematical shapes to develop a variety of forms, ground plans and layouts by transforming them, adding to them and subtracting sections from them.

> ◊

Geometrical planes

The geometrical properties of surfaces are largely transferred to the surface of the building. A square ground plan, which has four façades of equal length, is very suitable for designing what is basically a nondirectional building such as a pavilion in a park, or a building standing in an open space or a square. This applies even more to circular buildings, which are not only nondirectional, but also oriented to their own centre, thus heightening the significance of space. Rectangular buildings, by contrast, have their own orientation. Their unidirectional alignment establishes front and side façades. Along the side façades, they form distinct three-dimensional edges, whereas the front façades create stop ends. An elliptical building, on the other hand, forms an aligned space with two focal points. Like a circular building, it also has one continuous façade. In contrast to the circle, however, the ellipse (not unlike a rectangular building) provides orientation. If you develop ground plans based on these geometrical figures, you should be able to recognize these laws and apply them consciously. The principles described here are shown in the examples in Figure 46.

Geometrical bodies

By adding a third coordinate to the two-dimensional planes described above, you can create a variety of three-dimensional bodies. Geometrical bodies are subject to the same laws as surfaces.

Simple geometrical bodies are very striking and independent in character. They are especially suitable for designing object-like buildings

\\Tip:
You can perform rewarding experiments using simple working models of cardboard, plasticine, wood and polystyrene blocks to create the forms and proportions you want. You will find further information on working models and modelmaking materials in *Basics Modelbuilding* by Alexander Schilling, Birkhäuser Verlag, Basel 2007.

\\Hint:
Mathematics has played an important role in architecture ever since classical antiquity. Da Vinci's illustration of human proportions (1492) establishes a close relationship between the human body and geometrical forms. From time immemorial, architects have used basic geometrical forms to design ground plans, layouts, elevations and entire buildings.

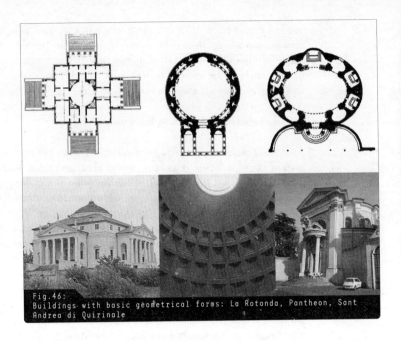

Fig. 46:
Buildings with basic geometrical forms: La Rotonda, Pantheon, Sant Andrea di Quirinale

set in spacious surroundings. › Figs 47, 48 Integrating such buildings into dense and heterogeneous settings where they stand close to other buildings is no easy task. The result is often a lack of clarity.

Forms found in nature

In addition to the material and immaterial relationships established by human beings, some forms and structures can be derived from nature, and they may give you ideas for your designs. Structures of this nature, which are often composed of organic shapes or forms derived from the environment, have captivating and richly varied contours. Most natural elements – even if they appear chaotic and randomly generated at first sight – are based on clear, albeit complex, patterns and rules. If you examine their cellular composition, you will discover structures that are unique with regard to their flexibility, loadbearing capacity, material efficiency and, above all, sculptural design. If you derive your ideas from nature, you will soon find inspiration in nature's formal language and diversity too.

Art Nouveau

The Art Nouveau buildings of the late nineteenth and early twentieth centuries were an expression of their designers' interest in nature. The diverse styles within Art Nouveau reveal a variety of ways in which architects were inspired by nature: from floral ornamentation to flowing, lively ground plans and façades to the organic design of entire buildings. › Fig. 49

Fig.47:
Right-angled cubes look compact in their surroundings

Fig.48:
Examples of the use of geometrical forms, cylinders and pyramids

Fig.49:
Art Nouveau forms inspired by nature

\\ Hint:
The term "bionics" is composed of the first syllable of the word biology and the second syllable of the word technics. Bionics endeavours to apply the method, structural and development principles of nature to technical systems.

Furthermore, studying nature can be very helpful if you wish to transform organizational and structural principles into constructed architecture.

Bionics

Many architects have developed or refined their building designs and structures by applying bionic principles. The tent-like cablenet structure designed by Frei Otto at the Olympic Park in Munich, the "bubble-and-foam" geodesic domes in Grimshaw's Eden Project, and Calatrava's bridges and buildings (whose ribs and loadbearing structures look like skeletons), have all been developed from structural principles that exist in nature. If you want to transform nature into technology, you will have to work your way through continuous processes of modification and abstraction, because natural forms are far too complex to be faithfully copied. Observing and analysing the principles upon which natural forms are based is far more important, for only then can the insights gained be transformed into architectural structures. Frei Otto developed roof forms by experimenting with steel cables and soap bubbles. The forms he created are produced by gravitational force alone and ideally reproduce the flows of force in the roof skin. › Figs 50, 51 and Chapter Working with materials and structures, Materials and structures as design elements

Plants and animals display a wide variety of features that can be adapted and transferred to buildings. Examples include the skeleton of a living creature, the compound eye of an insect, the shell of an armadillo and the wings of a bird. › Figs 52–55

Free forms

If you try to arrive at a design form by experimenting with structures and constructions you will be able to create forms that, at first sight, may seem to have nothing to do with order-creating systems and are fascinating for their sheer novelty. In its endeavour to create freely flowing forms, architecture comes very close to the other fine arts. This approach is particularly suitable for the design of large-scale building projects of social significance: churches, museums, cultural centres etc. Frequently, the goal of the design process is to make a building more dynamic and to render movement inside and around a building visible. Such strategies have been used in designing leaning structures with rounded-off edges or structures whose forms only become apparent during museum visits, or forms that spiral continuously through the building. › Figs 56 and Chapters Space and bodies and Arriving at ideas

Free forms are particularly useful for giving a building a symbolic character or lending a space or a building a unique appearance. The danger with this approach, however, is that it can lead architects to devote

Fig.50:
Baldachin spider's web on a meadow

Fig.51:
A roof made of a cablenet structure

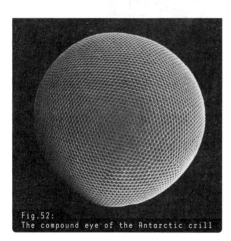
Fig.52:
The compound eye of the Antarctic crill

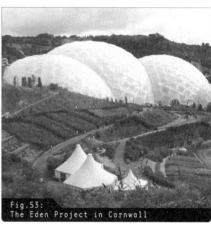
Fig.53:
The Eden Project in Cornwall

Fig.54:
A mute swan spreading its wings

Fig.55:
TGV railway station in Lyon

Fig. 56:
Dynamic free forms in modernist architecture

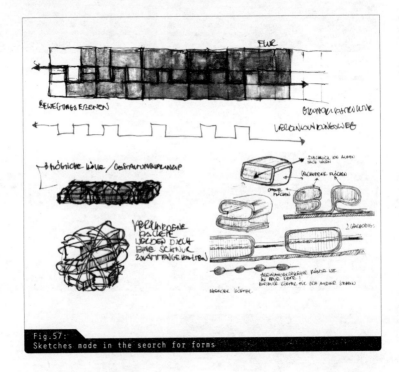

Fig. 57:
Sketches made in the search for forms

too much attention to the form and too little to the other demands placed on the building. › Fig. 57

SPACE AND BODIES

Architecture is always the result of interactions between spaces and bodies. Each body defines spaces; objects only become legible in space. These elements are opposites. They are also mutually and directly depend-

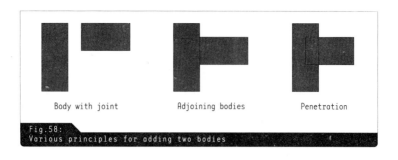

| Body with joint | Adjoining bodies | Penetration |

Fig. 58:
Various principles for adding two bodies

ent on one another. Only in their interaction do space and body form a whole. And architecture is possible only as a result of their interaction. Flowing successions of rooms often exist between clearly defined spaces and disbodies, with room sequences, gaps, and exterior and interior rooms emerging from the fertile interplay of both elements.

If you want your design to consist of more than just a basic form, you will find there are various ways of using a number of different of elements. You can create space by arranging and joining several elements. You can subtract bodies to create space or to modify and transform basic elements. There are endless ways of combining these methods.

Arranging and joining elements

If you want to create complex figures by means of addition, you must position the elements in your design in relationship to one another. You can choose a number of different solitary forms that interact through their form alone, or identical bodies that enter into a relationship through the way you arrange them.

Forms can interpenetrate or cut into one another, attach or be added to other forms, form rows or fit into one another. They always interact and create spatial structures. › Figs 58, 59 You have two basic choices here: you can either create space between two elements by arranging them in a certain way, or you can arrange the elements in a given space. › Fig. 60

\\ Tip:
The book *Architecture – Form, Space and Order* by Francis D.K. Ching is a standard work that explains the basic relationships between bodies and space (see Appendix, Literature).

57

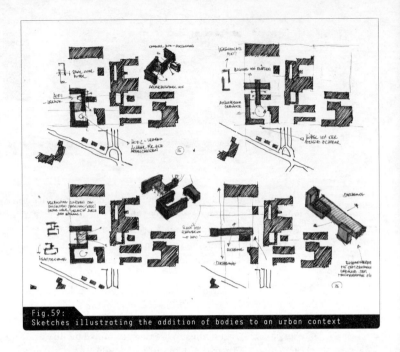

Fig.59:
Sketches illustrating the addition of bodies to an urban context

In addition to arranging two or three forms to create a composition, you will discover that there are situations (e.g. in urban and settlement development) in which a far greater number of forms must be coherently arranged. Typical arrangements include › Fig. 61

- Rows
- Grids
- Clusters or groups
- The arrangement of buildings round squares or other central points
- Radial arrangements
- Elements strung out like a pearl necklace

Creating space with slabs

The slab is a very special form of structure. It can be arranged horizontally or vertically. It can also be used to create paths and to steer people toward certain destinations. Depending on whether slabs are aligned to create a path or make a visual impact, they can be used to create spaces that flow into one another or interlock internal and external spaces. › Fig. 62

You should take into account not only the arrangement of the slabs on the ground plan, but also the impression created by the profile of the

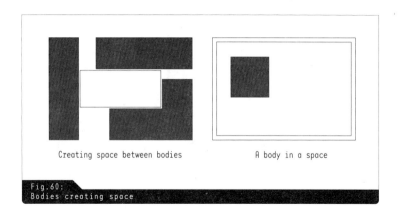

Fig.60:
Bodies creating space

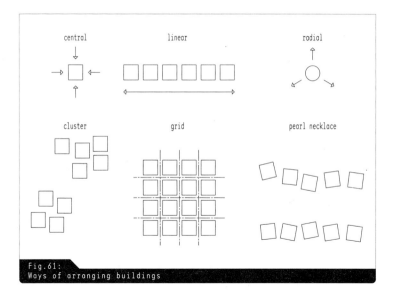

Fig.61:
Ways of arranging buildings

horizontal slabs, because it decisively influences the impact of the design. The vertical position of the base plate can play a vital role in determining a building's impact. It can be: › Figs 63, 64

_ Sunk into its foundations
_ Flush
_ Lying directly at ground level
_ Floating above it

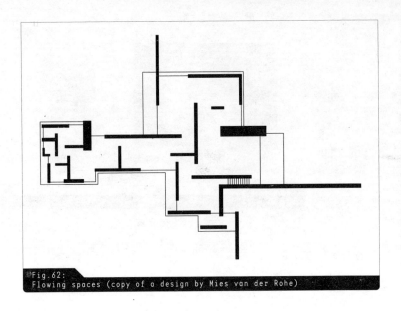
Fig.62:
Flowing spaces (copy of a design by Mies van der Rohe)

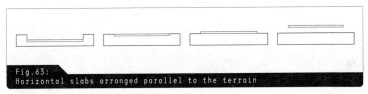
Fig.63:
Horizontal slabs arranged parallel to the terrain

Fig.64:
Examples of well-known Modernist slab houses

Subtraction and modification

You can develop surfaces and bodies by modifying or subtracting sections from them. For example, a fragmentary body can serve as the geometric basis for a design. Simple basic forms such as squares and circles can be varied and modified in countless ways. › Figs 65–67

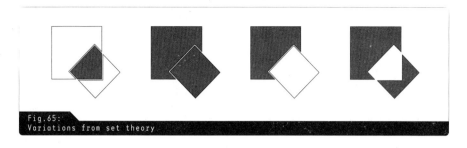

Fig.65:
Variations from set theory

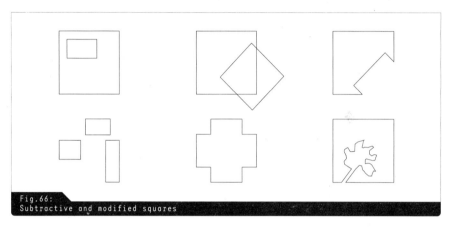

Fig.66:
Subtractive and modified squares

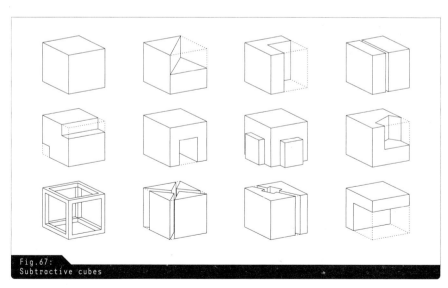

Fig.67:
Subtractive cubes

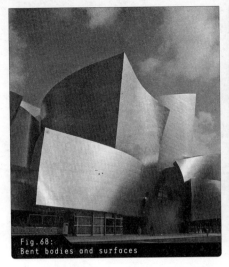
Fig. 68:
Bent bodies and surfaces

Fig. 69:
Folding as a design principle

Many of these variations are based on simple mathematical functions taken from set theory. If two bodies or surfaces intersect, they open up a number of possibilities, creating subsets, intersections, unions and differences.

Folding and bending

Another way of modifying or transforming bodies and surfaces is to fold or bend them. By bending or folding a strip, you can create a channelled spatial flow that contains and redirects space. By turning or bending a body, you can counteract its geometric rigidity, thereby making it softer and more ductile. Alternatively, you can change the basic form to take into account external influences and forces, which become apparent only when the form is transformed or deformed. › Figs 68, 69 In addition to transforming geometric elements by folding and bending them, you can also create and process far more complex forms, and freely create new designs using a computer.

WORKING WITH MATERIALS AND STRUCTURES

Historically, the starting point of design was the structure, which again came into its own at the dawn of the industrial age. In general, the architect will ensure that the structure, which is the building's basic frame, is visible in the final product. In skeleton structures, the structure may assume the form of a column grid that defines the façade system and the interior. Since the industrial revolution, engineered structures, such as bridges, have openly displayed their structural design, giving them a legible loadbearing structure. This development led at the end of the nineteenth century to the emergence of a new aesthetic that rejected the practice of cladding and decorating structures. Evidence of this can be seen in many areas of architecture too. In such cases, there is a direct and close relationship between the structure and materials used. Without the industrial production of iron, for example, filigree industrial buildings would be inconceivable. And without the properties of reinforced concrete, most shell structures could never have been constructed in their present form. ＞ Fig. 70 A material's technical, visual and tactile properties can be consciously used to add expression to a design and the ensuing structure.

MATERIALS AND STRUCTURES AS DESIGN ELEMENTS

The structure can be exploited to shape and structure the design. Alternatively, you can develop your design from the context or the function, or by creating your own forms and then integrating the required structure later on in a way that renders it imperceptible. In the case of buildings with demanding structural requirements (e.g. wide spans), it is very difficult to treat the function, form and structure separately. Hence, it may be advisable to take the loadbearing structure as the starting point of both the design process and all further developments.

Fig.70: A reinforced concrete hyperbolic paraboloid

If you want to use the structure as the basis for your design, you will have to familiarize yourself with the technical criteria for constructing loadbearing structures, i.e. the static systems, the choice of materials and the dimensions of the elements.

Structure and choice of materials

Generally speaking, the material from which a building is to be constructed is chosen during the design process. The material can, however, serve as the basis of the design. Natural stone may lend the surroundings their particular character, or the surrounding structures may be wooden buildings that were chosen for climatic reasons. > Chapter Designing in context, Social and socio-cultural factors The choice of material may be inspired by the original idea for the structure you wish to build. If your design is based on a specific material, you will have to take the properties of the material into account when you develop your structure or design. > Fig. 71

The structure as a design element

In general, the structure not only establishes order and supports loads, but in many cases also serves as the central feature of the design. The design ought to enhance both the loadbearing capacity of the structure and the use of the material. It can also decisively influence both the aesthetic appearance and the structure of a work of architecture.

Once you know the requirements and the static conditions, you can get down to the truly creative part of the work. There are many different ways of using loadbearing structures creatively. For example, distances can be spanned by linear girders, a loadbearing grid, flat building elements, cable supports or curved shells. Each of these loadbearing systems can determine the choice of design and the structure's spatial impact. Even if spatial considerations demand specific structural loadbearing systems, you can still purposefully use these systems to develop forms and structures that transcend their basic function as supports. You can

\\Hint:
Further tips on the choice of material and structure can be found in the following volumes in the Basics series:
Basics Materials by Manfred Hegger, Hans Drexler and Martin Zeumer
Basics Loadbearing Systems by Alfred Meistermann
Basics Timber Construction by Ludwig Steiger
Basics Masonry Construction by Nils Kummer

\\Tip:
The Spanish architect Santiago Calatrava has made many static models that serve as the basis for his often daring buildings. If you are interested in developing designs of this nature, you may find Calatrava's static experiments a good source of inspiration in developing your own ideas.

Fig.71:
Individual structures made of plastic, masonry, concrete and wood, which embody the specific properties of the materials

also design different types of connections such as hinged supports and erection joints intended for use with specific materials.

You need a sound knowledge of statics if you want to develop interesting structures that do not rely on standardized elements. The important thing here, however, is not whether you are able to calculate loadbearing structures precisely, but whether you have a good grasp of structures and static systems. You can also develop innovative ideas by experimenting with gravity. In doing so, you can develop internally stable static systems with the aid of small models, and save having to perform extensive calculations.

A structure that does justice to the material

You can obtain parameters for designs und forms by examining the structural rules and material properties of certain building materials. Masonry, for example, is of limited use for bridging wide gaps and is therefore often used in vaulted or arched structures. On the other hand, if you are constructing wooden buildings, you must consider the alignment of the material and take structural measures to protect the wood. Steel structures permit large spans with a minimal use of material, although

Fig.72:
Loadbearing structures that make excellent use of the materials from which they are made

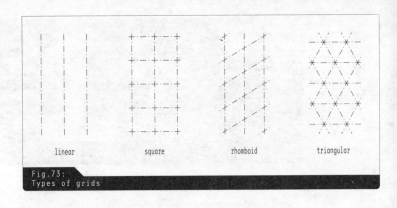

Fig. 73:
Types of grids

linear   square   rhomboid   triangular

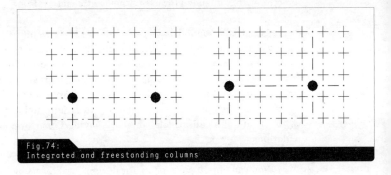

Fig. 74:
Integrated and freestanding columns

their fire protection behaviour is problematic. No matter what material you choose, the structure should take into account its specific properties so that you exploit them to the full. › Fig. 72

Modular grids and span lengths

Once you have chosen your material, its potential and the functional requirements will generally dictate the span lengths and the distances between the different structural elements in your design. A grid is created by repeating measured areas (if this is desired and forms the basis of the design). Modular grids are used in the development of organizational systems, façades and spaces.

Grids can consist of axes set at right angles to each other to form rectangular or quadratic fields. As an alternative, you might consider using grids with arbitrary angles and rhomboid fields, or triangular grids with equilateral triangles. The grid used in a linear loadbearing structure is based on linear spaces without a second directional axis. › Fig. 73

In this connection, the question usually arises as to whether the façades and non-bearing walls should rest on the grid axes or be offset against them. If the structure combines a skeleton and a solid structure in which the walls and the supports both assume loadbearing functions, it may be advisable to place the walls on the grid axes, thus breaking down the walls into rows of columns. By offsetting the walls and façade – as in a pure skeleton structure – against the grid, you will create a clear distinction between the loadbearing and the space-bordering elements. The structure, which will include some freestanding columns, will have far greater presence and appear more autonomous. > Fig. 74

HOW WE PERCEIVE MATERIALS

Materials are distinguished not only by their technical properties, but also by their effect on the viewer. We perceive materials in different ways through the interplay of our various sense organs.

Sensory perception

As almost 90 percent of all stimuli are perceived visually, the visual impact of buildings and materials has played an important role throughout the history of architecture, and has often been the object of study.

Materials and the atmosphere

It is quite difficult to capture and represent the sound, smell and tactile qualities of a building with the standard methods and tools of design. Such perceptions can be derived from their function or context only to a limited degree, because they are intimately connected to the material properties of a building. > Fig. 75

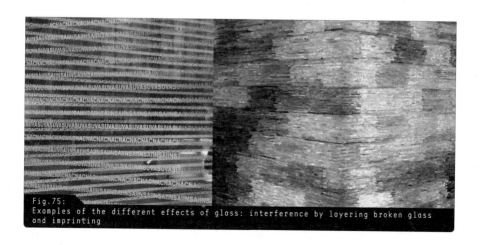

Fig.75:
Examples of the different effects of glass: interference by layering broken glass and imprinting

Fig.76:
Dematerialized floating glass bodies

Fig.77:
The old and the new in the contrast between stone and light structures

The skilful and correct use of materials can buttress your design statement or even be the very thing that makes it work. Attributes that are important in this context include:

_ Stony, earthy
_ Light, floating
_ Concealing
_ Layered
_ Transparent, translucent
_ Open, closed

With glass, you can more or less dematerialize a light body. Using earthy materials, on the other hand, you can enhance the structure of subtractive forms. If you use flexible, transparent building materials, you can create the impression that the shell is detaching itself from a building.
› Figs 76, 77

ARRIVING AT IDEAS

There are as many ways of arriving at an idea for a design as there are of practising architecture in general. The design process often starts unconsciously, or it may be inspired by an experience or an event that awakens your interest and stimulates your imagination. Such "events" are generally the various kinds of external impulses described above. They influence the architect creating the design or form the bases of experiences that he or she can draw on. Once you have developed your first ideas, you need to consider how you want to execute them and explore the potential of the different approaches.

<small>Remaining true to an ideal</small>

When working on a design, you will often discover that you cannot adhere strictly to your first idea from start to finish. Diverse parameters and basic ideas will necessitate a variety of options and design approaches. You may find inspiring ideas in different approaches and wish to pursue these further. There is, of course, the danger that you will water down good principles by combining them with others or by being too hesitant. You should keep your original principle in mind at all times, even if it means sacrificing other ideas. Abandoning a good idea in favour of another one is not always a disadvantage, because succinctness and legibility may enhance the quality of your design. When you start working on new projects, you may find yourself returning to ideas you had abandoned earlier. In this way, your ideas will broaden your knowledge.

<small>Simple but not banal</small>

Remaining true to an idea and searching for simple concepts also entails the risk of becoming banal or one-dimensional and sacrificing the complexity and subtlety of a design for the sake of clarity. Designs and buildings – just like the demands placed on a building – should be multi-dimensional and varied. However, variety also calls for complete clarity in the formulation of details. Many interesting designs are based on a clear principle that is faithfully executed down to the last detail. That said, your basic idea should not become a dogma. Nor should it mean reducing everything to this one principle. Make sure that the defining elements in your design articulate the design statement and do not detract from it. And do not forget that different principles can be allowed to overlap in different parts of the design so that they mutually supplement and enhance one another.

<small>Tools and techniques of presentation</small>

The presentation technique and tools you choose can greatly influence the design process. As in crafting an object, the result of your work depends on these tools.

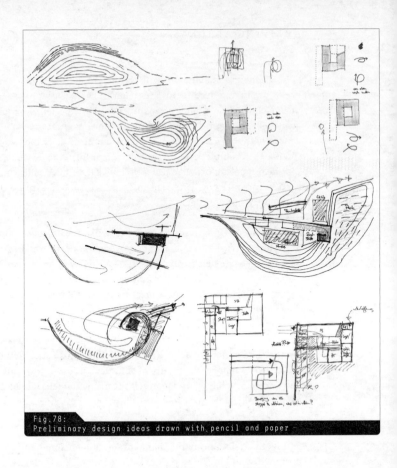

Fig. 78:
Preliminary design ideas drawn with pencil and paper

Pencils are very good for developing plans, sketching the ground plan, the framework, and the contours. > Fig. 78 The ground plan provides an ideal basis for studying and developing spatial structures and horizontal movements, as well as for organizing and zoning a design. It also allows you to represent vertical relations and spatial proportions in system sections. The façades provide a good starting point for developing and designing the room lighting, the building's interaction with the exterior and the outward appearance of the building. On the other hand, if you use media and presentation techniques that are both novel and unfamiliar, you may come up with some new and innovative solutions. > Chapter Arriving at ideas, Methods and strategies Over the past ten years, computer-aided design has made it possible to create building forms that would have been difficult to visualize with analogue tools such as pencils. One particular new trend – itself a product of the manifold possibilities of the digital world

Fig. 79:
Computer-generated forms (blobs)

– is called blob architecture, which uses free-form curves (splines) with complex, flowing and often rounded and biomorphic shapes. Only the use of modern design and visualization software has made such designs possible. › Fig. 79

Architectural students are recommended to try out various approaches so that they can find out which "manual" aids are best suited to intuitive approaches, and which ones best support and stimulate their creativity.

\\Tip:
Despite the potential of CAD programs, many architects find it difficult to design buildings with them because they cannot be used intuitively (in contrast to pencils and other tools). New students experience these programs in much the same way. For this reason, it is advisable to work with CAD software only if the manual side of design (e.g. working with a pencil) no longer absorbs a lot of your attention or if you have access to programs that allow you to design volumes and objects intuitively.

\\Hint:
If you are interested in background information on blobs and their development, we recommend the articles by Jeffrey Kipnis "Towards a New Architecture" and Greg Lyn "Architectural Curvilinearity: The Folded, the Pliant and the Supple" (both in *Theories and Manifestoes of Contemporary Architecture*, 2nd edition).

CREATIVITY AND CREATIVITY TECHNIQUES

When students first begin studying, they often wonder whether they will be able to fulfil the demands their college makes on their creativity as designers. They try to judge their own creativity and their design abilities on the basis of their previous experience, which they tend to project onto the architectural course. To what extent is originality necessary for an architectural design? What role does creativity play in the design process?

Creativity

When studying a design, you will rarely find yourself standing before a white canvas on which you can "paint" anything you like. There are certain parameters that influence a design process and often determine its points of departure and the directions it will take. During the design process, you will rarely develop ideas that have never existed before. The main challenge lies in developing new solutions based on existing principles and design approaches. Consequently, regardless of a person's individual potential, creativity should always be stimulated by external influences, the things that a person has already learned and the will to develop them further.

Science has developed a variety of creative techniques that are very useful in architectural design. These techniques aim to produce a large number of ideas intuitively in a short period of time. They trigger associations and new ways of thinking about problems. They also aim to activate hidden ideas and minimize inhibitions. Interaction between group participants is a very stimulating way of finding new solutions.

Brainstorming

In brainstorming sessions, a group is established and assigned a specific task. First, the task is explained and analysed. If required, a typical solution is presented. Then all the members are invited to submit spontaneous suggestions and ideas without criticizing one another or making judgmental statements about the others' ideas. Only in this way are group

> \\Hint:
> People generally associate the term creativity with a solution that solves a problem or a task in a novel, effective and, in most cases, unconventional way. It is quite possible that without creativity a specific problem would never be identified or solved. Furthermore, it helps you to approach a problem flexibly and with new resources that were previously unavailable in a specific context.

members discouraged from thinking through all the consequences of an idea first and encouraged to feel at liberty to express themselves freely. Even absurd ideas can motivate one member to come up with an alternative approach and thus encourage others to stimulate one another. All the proposals are noted down during the course of the session and then, when it draws to a close, presented and evaluated in terms of their feasibility.

Brainwriting

Brainwriting follows the same procedure as brainstorming, the difference being that the ideas are not submitted to the group, but written down by each member. This makes it easier for the more reserved members to present their ideas.

Gallery method

A technique often used to encourage creativity, especially in architectural courses, is called the gallery method. Each of the participants writes down his or her solution, which is then shown with the other solutions on a display surface. The solutions are discussed during an association phase (colloquium) and then refined on the basis of the new insights that have arisen. Afterward the revised solutions are presented for criticism by the other group members.

SCAMPER

SCAMPER is a creative technique that functions like a checklist. It aims to discover new directions and to question prevailing ideas. › Tab. 2 By answering questions about specific points, participants can try to find variations on designs and ways out of possible dead-ends.

Tab.2:
SCAMPER Method

Abbreviation	Meaning	
S	Substitute	Replace individual elements
C	Combine	Combine with other elements or combine different elements with one another
A	Adapt	Adapt content or functions
M	Modify	Modify the size or the scale; vary elements
P	Put	Put to other uses
E	Eliminate	Eliminate added elements, reduce to core function
R	Reserve	Turn upside down, diametrically opposed ideas

Mind maps

A mind map is a diagram presenting a task. The central task is written on the centre of a page and visually linked with subordinate aspects and variations. If, for example, you are trying to study a design assignment and its function, a mind map may help you to develop functional schemes.

› Fig. 80 and Chapter Design and function, Spatial allocation plan and internal organization

73

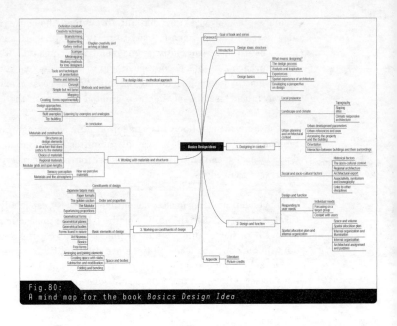

Fig. 80:
A mind map for the book *Basics Design Idea*

Working methods for lone designers

Even if you do not work in a group and are searching for a design idea on your own, you might still find the above-mentioned creativity techniques useful. To avoid the danger of becoming creatively blocked during the design process, it is a good idea to examine and question both yourself and your design as it take shape, and to consider these different aspects in a different light and from a different angle (e.g. that of an outsider). Discussions with others – including people working in other fields – can be helpful. You can also place yourself in the position of subsequent users and try to see the design through their eyes so that you can study a building's impact on them.

It is often advisable to take regular breaks and occupy yourself with other things for a while in order to put some distance between yourself and your design idea or the particular line of thought you are pursuing. This will allow you to challenge and critically evaluate the principles underlying your design through the eyes of an outsider.

METHODS AND STRATEGIES

Analyses and surveys

Usually, one of the very first tasks you face involves examining the above-mentioned design parameters. There is no way of knowing which approaches will be successful. It may be best to visit the site. While you are making sketches, taking photographs and studying the wider surroundings, you will develop you first ideas for your design. Analysing the spatial and urban context in order to get a feel for an area is often a worthwhile endeavour. This is particularly important if you are dealing with an area unfamiliar to you.

Models and built examples

Another method is to study other buildings designed to solve similar tasks or conceived in a similar context, so that you can get a feel for the assignment. If the planning assignment involves a special function (such as a railway station, or a building site located on a slope) that makes it difficult to survey the surroundings, examining examples of constructed buildings can give you useful clues on how to solve the problem. The more you study examples of constructed buildings, the greater the range of variations and choices available to you will be. When developing your design, you should remember that aesthetic sensibilities are not the only criteria that count. You should also take into account the general conditions, try to visualize and analyse a building or vision in its specific context and develop your own ideas and approach accordingly. › **Chapter Designing in context**

Design approaches of famous architects

Another method is to study the works of a well-known architect and to try to develop a design by applying his or her guiding principles and ideas. While you are designing your project, you will gain a much deeper understanding of the architect's work and standpoint. By applying the contents to one of your own designs, you will be able to relate to it and develop your own approach as you see fit.

Theme and leitmotif

Frequently, buildings do not embody an identifiable design idea. This is often due to the great number of functions buildings are supposed to perform and to the fact that these functions often have to be given more or less equal treatment during the design phase. To ensure that your design has an identity, you might assign it a theme or leitmotif, or find a good foundation. Sometimes, leitmotifs can be found outside the normal context of architectural design.

You may find inspiration for themes from both the material objects and immaterial phenomena that you transform into elements of your design. The difficulty lies in correctly defining the degree of abstraction that this transformation entails. If you select a theme that is distinguished by very explicit motifs and associations, make sure that your design does not appear simplistic or banal because you have directly adopted the motif in your design. At the same time, you should also ensure that your motif does not lose all relevance to the original theme by appearing overly abstract.

Fig.81:
Realizations of the "ship" leitmotif in different epochs

Fig.82:
The "evolution" leitmotif, interpreted as a spiral that museum visitors walk through parallel to the thematic exhibits

Fig.83:
The "island" leitmotif for the design of a housing estate in the forest

Your design should respond subtly to thematic influences. > Figs 81–83 These exercises will help you learn how to use referential motifs, abstractions and symbolic elements.

Mapping

Mapping is a special form of transcription and a particular way of developing a design theme. It allows you to develop an architectural structure that relates to spatial phenomena at the future construction site and serves as a basis for your design. The actual process of mapping can involve taking photographs and videos or making sketches and models. The data are then evaluated using a specially developed logic and a language

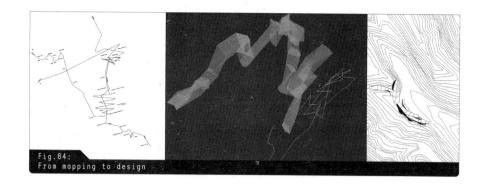

Fig. 84:
From mapping to design

– a special system of notation – that concentrates on spatial phenomena (e.g. spaces for movement, light and shadow, gaps, noises etc.) The notation is designed to translate spatial phenomena into a model architectural structure. › Fig. 84 This highly intense and undoubtedly subjective analysis of the property will allow you to study the site closely and prepare yourself more effectively for the task at hand.

Creating forms experimentally

Instead of approaching the task analytically, you can also arrive at an initial thematic position by creating forms experimentally. You can perform this task on a computer using design and visualization software or by using analogue, sculptural models (e.g. gypsum, wax, plasticine). You can digitize your results later using a three-dimensional scanner or other equipment. By experimenting with traditional materials from different contexts, you can create unanticipated, innovative forms.

\\Hint:
The term "mapping" is derived from the word map. Maps deal with facts. They are graphic representations of dimensions, attributes and relationships between elements in the physical world or in the world of logic. Almost everything can be illustrated or recorded: spaces, galaxies, time, history, occupations and philosophies. If you are interested in this subject, we recommend *Mapping* by Roger Fawcett-Tang and William Owe, RotoVision, 2005).

Fig. 85:
Constructed examples based on forms created experimentally.

The many available methods include:

_ Deriving a form from crumpled-up paper (Frank O. Gehry does this when he is preparing his studies)
_ Deforming an object (e.g. a can or a cardboard box) to create complex geometric forms
_ Creating structures using magnetism: e.g. arranging iron filings or nails on a table
_ Processing plasticine on a board to create dynamic forms

Experimental work may initially seem to have an accidental or random quality, but this is only true to a certain extent. You will generally find yourself experimenting with ideas and discarding results until you know you have finally got a structure with potential. You can then set about refining and optimizing it. In this way, the experiment serves as a catalyst for your creativity. The next stages of the design process proceed – as with basic geometric forms – from this first move. The exciting thing is that, whatever approach you choose, you can identify the functional and aesthetic potential in the forms you have created. › Fig. 85

IN CONCLUSION

Although the quality of an architectural design is manifested in certain characteristic features such as the consistency of the concept and the differentiated approach to the site, judgements on quality always include subjective perception. Architectural quality cannot really be judged in terms of objective parameters or scales. An evaluation always includes the subjective views of the evaluator. Periodic changes in taste beyond the lifetime of a building also play a role. Buildings that were modern in the 1980s may now appear outdated, whilst late nineteenth-century Central European (*Gründerzeit*) buildings overladen with stucco are currently very popular, despite the derision poured on them by the Modernists for being kitschy. Regardless of whether a design is made to be "timeless" or whether it directly reflects contemporary tastes, it will only be truly consistent if it is well conceived and executed without any loss in quality. The design process evolves out of an initial idea; the design gradually takes shape. The aim now is to transfer the principles and the basic idea underlying the design into the plans; to work out the details; and to steer the project through to the construction stage.

The present book is intended to be a source of inspiration for designers – in full awareness of the fact that it merely goes through the preliminary phase of a process. The methods presented indicate various – and by no means definitive – ways of tackling design assignments. Design does not involve simply reproducing certain processes or existing models. It is always a creative process that produces new totalities under a variety of conditions and draws on diverse sources of inspiration. The elements that we have examined independently in this book to come up with our first design ideas – context, function, design, material and structure – must interact closely throughout the design process to create a complex system of influences and conditions. Which of these elements gives birth to the first design idea and which approach leads to a mature concept? The answer lies in individual experience and insights. What is more, a course in architecture provides an invaluable basis for gaining them. Design is not something you can learn passively from books or lectures, which merely provide stimuli or act as catalysts for students' individual development. Design is something you learn by doing – through practical experience. I hope that the book *Design Ideas* provides useful stimuli for making your own creative designs and encourages you to devote yourself to every aspect of design. Above all, I hope that it will help you, in the long term, to find your own path in design. This means that you must never stop asking questions, and that you will always be willing to experiment, that you retain your curiosity and find pleasure in working on your own designs.

APPENDIX

LITERATURE

Design basics:
Leon Battista Alberti: *The ten Books of Architecture:* the 1755 Leoni Edition, New York 1986
Francis D.K. Ching: *Architecture. Form, Space and Order*, John Wiley & Sons, New York 1996
Le Corbusier: *The Modulor, Modulor 2*, Birkhäuser Verlag, Basel 2000
Roger Fawcett-Tang, William Owen: *Mapping*, RotoVision, Brighton 2005
Christian Gänshirt: *Werkzeuge für Ideen*, Birkhäuser Verlag, Basel 2007
Jeffrey Kipnis: InFormation/DeFormation, in *Arch+*, No. 131
Greg Lynn: Das Gefaltete, das Biegsame und das Geschmeidige, in *Arch+*, No. 131
Andrea Palladio: *The four Books on Architecture*, MIT Press, Cambridge 1997
Camillo Sitte: *City Planning according to artistic Principles*, Phaidon, Berlin 1965
Vitruvius: *Ten Books on Architecture*, Cambridge University Press, Cambridge 1999

Architectural history and theory:
Otl Aicher: *The World as Design*, Ernst, 1994
Leonardo Benevolo: *The European City*, Blackwell, Oxford 1993
Le Corbusier: *Towards a new Architecture*, Dover Publications, London 1986
Siegfried Giedion: *Space, Time and Architecture: The Growth of a new Tradition*, Harvard University Press 2003
Hanno-Walter Kruft: *A History of architectural Theory from Vitruvius to the Present*, Zwemmer, London 1994
Robert Venturi: *Complexity and Contradiction in Architecture*, Little Brown & Co 1977
Robert Venturi, Denise Scott Brown, Steven Izenour: *Learning from Las Vegas*, MIT Press, Boston 1972

Different focuses on design:
Jürgen Adam, Katharina Hausmann, Frank Jüttner: *Entwurfsatlas Industriebau*, Birkhäuser Verlag, Basel 2004
Sophia und Stefan Behling: *Solar Power*, Prestel Publishing, Munich 2000
Mark Dudek: *Entwurfsatlas Schulen und Kindergärten*, Birkhäuser Verlag, Basel 2006

Roberto Gonzalo, Karl J. Habermann: *Energy-Efficient Architecture*, Birkhäuser Verlag, Basel 2006

Rainer Hascher, Simone Jeska, Birgit Klauck: *Office Buildings. A Design Manual*, Birkhäuser Verlag, Basel 2002

Paul von Naredi-Rainer: *Entwurfsatlas Museumsbau*, Birkhäuser Verlag, Basel 2004

Ernst Neufert: *Architect's Data*, Blackwell Science Ltd, London 2000

Friedericke Schneider (ed.): *Floor Plan Manual. Housing*, Birkhäuser Verlag, Basel 2004

PICTURE CREDITS

Figures 1, 3, 5 (W. Wassef), 12 left (R. Moneo), 26 right (E. Heerich), 30, 34 left + centre, 35 all, 46, 49 left (H. Guimard), 70 (H. Stubbins), 71 right (T. Herzog), 77 (M. Campi, F. Pessin), 80	Sebastian El khouli
Figures 2 (Morger Degelo), 46 centre, 75 left (HdM)	FG ee, Hegger
Figure page 10 (Le Corbusier), Figures 4, 6, 11, 12 right, 13, 14, 18, 20 (Botta, Palladio), 23, 24, 26 centre left, centre right (Gaudí), 27, 29 right, 33 (Foster), 36 (Le Corbusier), 39, 41, 47 all, 48 all, 49 centre left + right (Gaudí), 55, 56 left (Mendelsohn), 56 right (Le Corbusier), 58, 60–63, 64 left (Mies van der Rohe), 65–67, 71 centre left (Gaudí), 71 centre right (Le Corbusier), 73, 74, 75 right, 81 all, 85 centre right (Domenig)	Bert Bielefeld
Figure 7 left	Katrin Kühn/ Sonja Orzikowski
Figures 7 right, 21	Barbara Gehrung
Figures 8, 15, 25, 57, 59	Isabella Skiba
Figures 9, 19	Rahel Züger
Figures 7 centre, 10, 16, 17, 26 left, 34 right, 40, 83	Atelier 5
Figure 22	Joost Hartwig, Nikola Mahal
Figures 28, 29 left, 78	Annette Gref
Figure 31 (Atelier 5)	Leonardo Bezzola
Figure 32 (Brüning Klapp Rein)	Brüning Klapp Rein

	Architects
Figures 37, 69, 84 (DGJ)	Hans Drexler
Figure 38 (R. Serra)	Andrew Dunn
Figure 45 (Le Corbusier)	FLC, ProLitteris, Zurich
Figure 49 centre right (G. Strauven)	Peter Clericuzio
Figures 50, 51 (Frei Otto), 53 (Grimshaw), 64 right (M. van der Rohe), 71 left, 72 left (Calatrava), 72 centre (Frei Otto)	Free pictures
Figure 52	Gerd Alberti, Uwe Kils
Figure 54	arp
Figure 56 centre (Utzon)	Denn
Figure 68 (F. Gehry)	Jon Sullivan
Figure 72 right (Nervi)	I have got the style
Figure 76 (Behnisch)	Chris73
Figure 79 (P. Cook, C. Fournier)	Marion Schneider, Christoph Aistleitner
Figure 82 (Zamp Kelp/Krauss/Brandlhuber)	Cordula
Figure 85 left (Z. Hadid)	Richard Wasenegger
Figure 85 centre left (Coop Himmelblau)	Andreas Pöschek
Figure 85 right (Gehry)	Cacophony
Figure page 80	Martin Zeumer

Every effort was made to acknowledge and obtain permission for all pictures. Please inform the publisher of any mistakes or oversights that might have occurred.

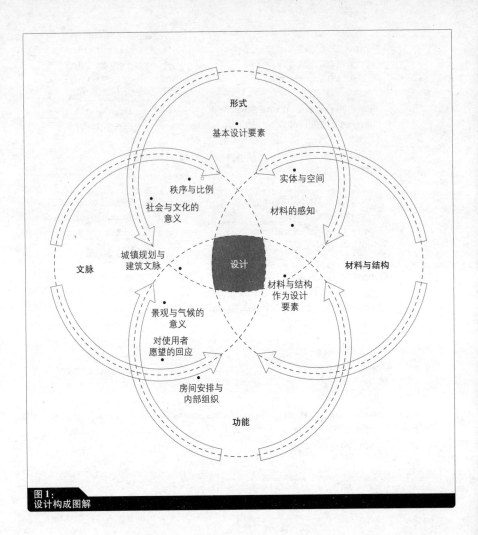

图1：
设计构成图解

导　言

　　建筑不是在真空中创造出来的，它通常要对已有文脉进行回应，并使之最终成为文脉的一部分。建筑也要承担一定的功能，对某种任务提供切实可行的解决方案，并且通过设计和材料使之成为现实。从"设计构成图解"中可以看出，图中所设定的文脉、功能、形式、材料和结构等变量与每一个建筑设计都是直接相关联的（见图1）。这些变量也是建筑设计涉及的各种要素。不但如此，当把这些要素作为设计概念的发展策略时，将具有更大的潜力。

　　下列章节将系统地介绍与设计相关的各个变量，并从方法与理念的角度对其进行分析。各种不同的其他设计关联因子的连接将被确定，以强调其连接方式和从属关系。这些互相参照的变量表明，个别主题之间是怎样一种相互纠缠的关系，并且也使你在创意阶段避免陷入僵局。另外，各章节还给出了一些建筑实例以及进一步阅读的材料，使你能够对所讨论的方法论问题及其在建筑设计上的应用进行更深入的研究。

　　这些设计变量形成的框架，你可以在自己的设计过程中尝试应用。这些变量使你在设计初始阶段以一种结构化的方式触及信息和灵感的相关来源。在设计过程最开始的时候，对所有已知信息、条件、感受进行罗列，并把它们以一种连贯的方式进行视觉化往往是很有帮助的。这种联系经常会揭露出一些不引人注意的关系和要点，而指出存在于知识结构中的盲点以及可能的矛盾。

　　本书最后的一部分介绍了不同的设计方法和练习题，帮助你跨出第一步，通常也是最困难的一步，进入设计的程序。它所关注的是与你的设计工作相关联的那些个别的节点。

设计基础

设计过程

设计是一个复杂的、充满矛盾的、非线性的过程。这无论对于有工作经验的建筑师还是初学者都莫不如此，因为这是设计的本质所在。即使设计任务要求的细节是清楚的，设计过程和目标也是未知的。学习设计就是去着手探求一种方法，这种方法能够去确认联系点和所属关系，能够去理解任何给定设计任务的参照体系。然后，建筑师运用他们的知识、经验、空间想像力和创造性，把这些要素转化成建筑。

任何设计都会引出新的问题，使你有机会获取新知并创造出适合设计任务的新的原型。设计活动在建筑师的职业中不但是联系所有一切的核心元素，它也是这个职业中最有意思的一个方面。

提问而非回答

当一个新的设计任务开始的时候，非常重要的是，要提出正确的问题，而不要急于去寻求简单的答案。这些显而易见的答案与设计任务的复杂性很可能是不相称的。大量的设计问题会在具体的设计任务中浮现出来，对于特定设计条件的分析或者对建筑实例的研究可以作为趋近设计任务的有效途径，你可以在各种策略和方法中进行选择。

分析与灵感

一个普遍的方法包含着对最重要变量的详细研究与分析：
— 城市规划条件/景观条件；
— 地段的历史沿革；
— 使用者/使用需求；
— 有着类似条件和类似功能的其他建筑。

将这些信息与分析的结果联系起来会有助于你产生一些想法，这些想法会导向一个具体的设计概念。除了要进行科学的分析，你可以寻求其他更有意思的方法，因为它们没有什么限制，这些方法能够带给你更大的自由度（见"设计概念的形成"）。

另一个途径是在设计过程的最开始去寻求灵感或想法。想法可以来自设计任务的个别的细节、要求，或者来自于并非与设计任务直接相关的某些启示（见"设计概念的形成"之"方法与策略"）。随着设计工作的进展，其他的要求和设计阶段会逐渐融入设计的创意之中。于是，设计在一个持续变化的过程中得以发展。

方法选择的正确与否取决于个人的工作风格、技巧以及实际的设计任务。每个设计的情况都是不同的。所有的学生在他们的课程学习中都应该利用这个机会尝试不同的途径和解决方案，目的是去认识各种方法的优势和弱点，去发现那些最适合他们自己的方法。

体验

个人体验和感受在提出创意的过程中是决定性的。在每一个练习中，你可以磨砺你的设计工具，开发你的设计感觉。使用钢笔、电脑和模型进行操作仅仅是通往终点的途径，不断练习的最重要的报偿来自于智力的层面。当你跳出惯常思维的误区，当你提出新的想法并且通过"试错法"进行设计时，你就摸到了创意之门并且发展出与众不同的建筑技能。对创造性的开发并非终止于大学教育，而是一个需要终生倾力投入的过程。

那些并非直接与设计任务相关的外部影响在设计过程中也起着决定性作用。当设计工作以团队的方式进行时，设计概念产生自与他人的对话、产生于每位团队成员对设计过程的贡献、产生于对他人的建议，最终找到正确的设计途径。这种情况同样适用于建筑师与委托人的互动，或者大学里老师的评图。交流想法有助于突破个体思维的局限性，集中的外部反馈可以为个体提供持续的推动力，避免过早认输。学生们从中学习某些方法是否有助于设计目标的实现，他们从他人的经验中获益（见"创造性与创新技巧"）。

建筑的空间体验

建成的建筑同样能提供丰富的经验。对建筑的深入研究和切身体验也是一个学习创意与方法的好途径（见"设计概念的形成"）。尽管书籍也能够为学生介绍新的领域，在学习过程中充当了他们灵感的源泉，但是书籍内容毕竟有一定选择性，不能传达出建筑全方位的文脉关系。当学生用所有的感官去感知建筑时，将获得长久的记忆和重要的体验。参观建筑最根本的是去体验它的空间、从其周围各个角度去观察。最根本的是去触摸和感觉建筑、去观察人们是如何使用它的（见图2）。对于学生来说，这是获取建筑的全面概念惟一的方法，也是提高自己在工作中的洞察力的惟一方法。只有当他们亲身体验建筑时，它们才能对自己的工作产生持续的影响（见图3）。

> **注释：**
> 你要尽可能多地去看建成的建筑。从你所在的城市入手，仔细观察和分析那些商业街或居住社区中的建筑，这将使你对环境有个感觉。同样重要的是去研究那些出自著名建筑师之手的建筑。在城市中漫游、走走停停、在学习期间或之后进行远足，利用所有这些机会去看那些你所在城市或区域中的建筑名作。

图2：
建筑的外观不但受初始设计的影响，同样也受后来的住户的影响

图3：
现场参观可以确认居民是否能够接受建筑的初始设计意图

形成一种设计观

你所经历的设计任务以及你对已建成建筑的思考将使你逐渐形成一个如何去解决设计任务的观点。"设计观"（design perspective）是指一种接近设计的意识，一种用来将设计转变为现实存在的方法。这并不需要在建筑语言的个性方面追求某种标新立异的风格，相反，设计观是一个作品的统一性原则，它是你处理设计任务和项目所采用的方法的直接产物。

设计观往往与设计者的个性直接相关，而且并非仅限于和建筑的相互作用。它可以是一种与更广泛的社会或哲学的背景相关联的、完全的个人世界观的表现。因此，设计观的形成是个性成熟过程的一部分，不能生搬硬套。当建筑学生开始探究著名建筑师的建筑美学观和设计观时，他们容易去寻找那些清晰可辨的样板和方法，以及那些可以在自己的设计中派上用场的东西。他们自然而然地发现，去理解那些已有的方法和观念、在学校的设计课上去尝试应用从而获得某些经验是很有帮助的。这是一条探索已知设计奥秘的惟一途径。然而，那些著名的建筑师不会用任何单一的教条束缚自己，不会限制自己的发展，不会把自己局限在某一个特定的方法上。

设计与文脉

每个设计都是在特定文脉中出现的，无论是建筑环境，或者是社

图4：
观察公共广场的人流规律

图5：
光线随时光偏移，阴影遮蔽着一个安静的角落

会的、文化的环境。

虽然设计的过程开始于对场地的研究，设计的结果却并不总是必须与场地周围的条件相协调。一个独立的姿态，或者反抗的姿态也不失为一种选择。尽管如此，为了理解某些决策的影响效果，仔细地研究场地条件是非常重要的。根据场地所处的乡村或城市环境的不同，自然的或人文的影响具有支配性的作用。

当地条件

在大多数情况下，对场地及其周边的深入研究对于设计概念的形成是至关重要的。你可以尝试通过速写、测量、踏勘等手段从三维尺度去把握场地，特别是当场地的地形特征比较特殊时更应如此。你还应当给自己充足的时间去研究场地周围的视线关系和景观要素。

注释：

建议在不同的时间去参观建筑现场、观察日常生活。行人都在哪里行走？他们从哪个角度观察场地？安静的区域在哪里？街道的噪声来自哪个方向？场地的气氛是如何形成的？一天之内光线是如何变化的(见图4、图5)？

图 6:
景观模型：乡村环境中的建筑

图 7:
景观模型：城市环境中的建筑

景观模拟　　如果需要考虑更广泛的景观因素，或者场地的地形特征比较特殊时，就需要制作一个景观模型以表达高差关系。这种模型可以用来检验和优化初始设计方案对于周围空间的影响效果。同样重要的是，可以利用模型研究从各种角度观看场地的效果（见图6）。如果更大的文脉需要被描述——城市系统、建筑物之间的关系等等——所有这些重要的关系都要结合到模型之中（见图7）。在城市环境下当你准备着手一个设计时，为了找到场所感，你也要对场地本身及其周围的环境进行空间分析。这些分析可以采取的形式是多种多样的，如拟建图（as-built plan）、开发结构、街道与通道间的联系、广场、绿地的设计等（见图8）。除了上述的优势，景观模型使得我们能够从一定距离观察场地的各个方面，从而发现那些通常从场地本身不易发现的关系。

　　对于场地的研究有助于你理解潜藏在现场背后的无形规则。各种系统、各种要素之间的关系被关注并且凝结成一个结构，成为一个设计的根基。设计可以和谐地结合到这个结构中，也可以采用另类的方法去演绎这个结构。同样地，你可以故意地去选择与周围环境发生

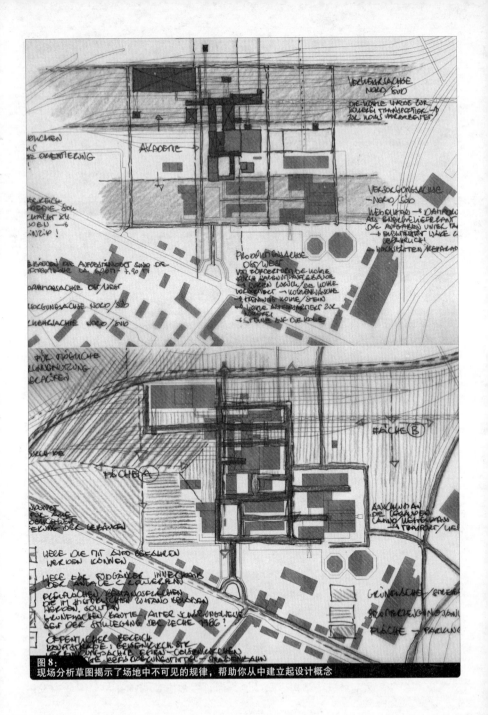

图8：
现场分析草图揭示了场地中不可见的规律，帮助你从中建立起设计概念

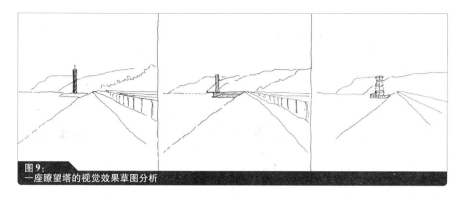

图9：
一座瞭望塔的视觉效果草图分析

"冲突"，或者去发展一种自治性的（autonomous）设计概念。最根本的是，设计作品要建立在对场所整体性的理解之上。如果你最终寻求一种与周围环境的对立，就必须是有意识的选择，而且必须是容易理解的。

景观与气候

对超越场地的更大范围的景观研究可以产出各种不同的设计策略，既可以设计一个独特的地标性建筑，又可以尽量简化可见的建筑物，使之融入周围环境之中。

地形　　场地地形决定了设计与景观结合的方式。不论场地是平地还是坡地、台地还是山地，地形总是影响着建筑以及建筑室内外的空间关系。地势地貌同样也影响着建筑内部空间的布局（见图10）。例如，景观中的高差可以延续到建筑内部，入口可以布置在与公共街道空间相联系的通道上（见"设计与文脉"之"城市规划与建筑文脉"）。当然，如果设计的某些部分允许的话，场地也可以被当作景观来设计。总之，建筑物的设计可以反映地形特征，或者干脆成为地形的一部分。然而，建筑师也可以有意识地决定不采用上述方法，而去创造一个自足的、独立于地形的功能单元，在建筑与周围环境之间划一条明确的界限。

坡地　　当场地内的高差大于一层时，你必须考虑建筑的内部结构如何去适应地形的状况，必须考虑在室内空间和外部环境之间应当建立怎样的关系。

坡地上的建筑可以置入坡地，也可以悬挑于坡地之上，或者随坡就势处理成台阶，甚至对坡地进行改造（见图11）。这些不同的形式在室内与室外空间之间形成了不同的关系。地形的特征常常会启发出有

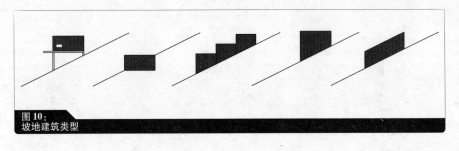

图 10：
坡地建筑类型

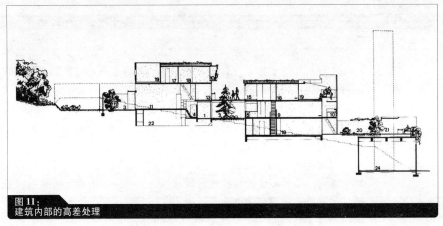

图 11：
建筑内部的高差处理

魅力的设计概念，特别是在场地复杂的情况下更会如此（见图12）。如果在场地与周围景观之间有宽阔的视野，你必须决定建筑在风景中展现的角度以及那些在风景中可能会引起兴趣的关系（见"设计与文脉"之"城市规划与建筑文脉"）。

反映气候特征的建筑

除了场地地形，对气候条件的分析为设计的发展提供了进一步的方法。根据地域性气候条件，建议以太阳来确定建筑的朝向和布局。建筑可以向朝阳的一面开敞，让更多的太阳能量进入并储存在建筑内部，或者向太阳关闭，将热量排斥在建筑之外。

此外，可以通过对建造方法、材料和建筑形式的调试以适应某个地区的宏观与微观气候条件。在炎热地区，建筑物可以用较重的材料建造，或者建于地下，以利用建筑材料或土壤的蓄热能力从而达到冷却的目的。或者，建筑师也可以通过利用主导风向组织穿堂风。相反，在温和气候地区，可以通过优化建筑物的体形系数以减少传导热散失，增大立面的日照受光面积。

图12：
景观作为设计的基本要素

城市规划与建筑文脉

在人文环境——即由人类所创造的环境——之中，人类活动的因素通常会比自然因素对设计施加更大的影响。人文因素可以通过对环境的详细研究来分析，其目的是建立可能的设计途径。场地周围环境通常包括相邻建筑物、街道或者树木所形成的参照点。场地内也可能有其他建筑物需要你在设计方案中结合进去。如果有建筑物邻接你的场地，你还需要考虑你设计的建筑是否可以（或者必须）紧邻而建。在密集的城区，建筑物临街而建的情况下，你必须还要考虑你设计的建筑是否要采用与相邻建筑相同的形式。如果建筑处在联排建筑的夹缝中，就必须对相邻建筑的不同高度作出回应（见图13）。如果建筑作为独立的结构矗立在街道上，你可以使之适应各种既有体系，或者为既有体系提供一个对应物。在这种类型的分析中典型的参数包括：

— 屋顶形式；
— 建筑朝向；
— 街道与建筑之间的距离；
— 材料；
— 窗户的形状和尺寸。

与原有建筑相呼应

在将来，大多数建筑任务将不再以在未开发的土地上建造新的建筑物为中心。甚至现在，为如何利用既有建筑而寻求创意和设计的需求在日益增长。这种设计任务给你提供了一个机会，考虑在新老建筑之间存在的差别，提出新的设计策略。你可以赋予老建筑以新用途，使之适应现代生活的需要，对之进行某种改变，使之具有全新的特

图13：
建筑的屋顶影响着城市环境的面貌

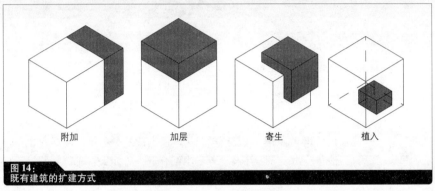

附加　　　加层　　　寄生　　　植入

图14：
既有建筑的扩建方式

性。你可以保持老建筑原貌，或者将之纳入新的发展规划（历史性地标建筑、功能可变的建筑等），甚至利用个别的建筑元素（立面、结构等）(见图14)。另一种选择是，如果改变现有建筑的功能代价太高，就可以完全拆除。在这一过程中，新建筑介入的目的是至关重要的问题：新与旧是应该各自独立存在成为两个截然不同的部分？还是二者之间应该有对话？建筑师是应该努力追求项目的广泛一致性？还是应该强调差异化的质量(见图15)？从经济、环境和文化的观点来看，去追问拆除的合法性或必要性总是十分重要的(见"设计与文脉"之"社会与文化因素")。

城市发展条件　　为了避免无序的增长以及对历史建筑的破坏，很多国家都为房地产的开发制定了详细的法规。这些法规包括：

—— 限制或规范建筑密度或建筑面积；
—— 控制建筑的边线；
—— 规定建筑层数；

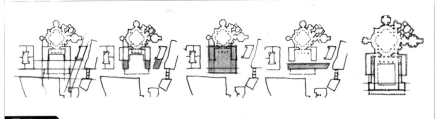

图15：
草图分析既有建筑的扩展方案

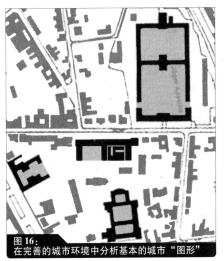

图16：
在完善的城市环境中分析基本的城市"图形"

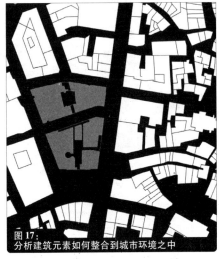

图17：
分析建筑元素如何整合到城市环境之中

— 规范屋顶形式和施工方法；
— 规定入口位置和交通组织方式；
— 规定与其他建筑或用地边界的距离；
— 保护古树名木。

假如你正在设计一幢将要被建成的建筑，这些法规通常是必须要遵守的。而且，在方案的早期还需要对城市规划的条件进行研究，以确保方案能够最终实施。但是，如果这些法规非常严格，没有回旋余地，那么就存在一种风险，即：对法规的熟悉可能阻碍设计过程中创造性的发挥。在设计概念的自由生成与可实施性之间寻找一块中间地带是非常重要的。这将决定你的研究所需要的细节以及你的分析所要达到的深度。

城市参照物与轴线

　　密集紧凑的城市环境通常会有一些参照点或参照建筑,这些参照物作为主导性的城市设计原则,决定了周边区域的大多数设计(见图16、图17)。作为城市规划分析的一部分,你应该去研究你所设计的建筑其功能在更大的城市结构以及在场地周边环境中所扮演的角色、所发挥的作用。你也可以通过此研究得出一个基本的城市"图形"(figures),以此作为深入分析的出发点。

　　例如,如果相邻建筑退后于街道边线,你便可以设计悬挑的建筑以创造一种城市氛围。另一种选择是,通过设计后退于街道的建筑,你可以获得一个前院或者庭院空间。如果周围地区已有大

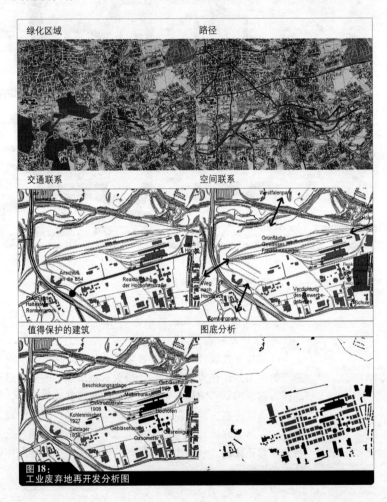

图18:
工业废弃地再开发分析图

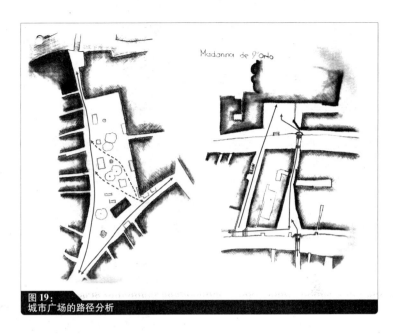

图19：
城市广场的路径分析

量主导性的要素，那些轴线和形式会对你的设计产生影响，那么这幢建筑就会很好地整合到城市环境之中，并在城市中表现出一种谦虚姿态。

根据不同的周边环境，重要的参照点或者独立的建筑可以为尺度关系和轴线提供一个基础：一个非对称的广场创造了多角的界面，而场地对面正交的街道为建筑打开了一个观察角度。你还可以为城市空间中重要的元素创造一个对应物（counterpart）。通过设计与城市环境对话的方式是多种多样的。

注释：
除了要研究当地状况，你可以利用区域地图、城市地图或者土地产权登记图进行各种分析（见图18）。拟建图把周围建筑涂成黑色方块，可以把你的注意力引向城市建筑的肌理和空间，而街道地图则可以揭示出重要的空间联系（见图19）。

提示：
更多关于设计与公共广场空间类型方面的研究可参见《城市规划的艺术性原则》（City Planning According to Artistic Principles），卡米洛·西特（Camillo Sitte）著，哥伦比亚大学出版社（Columbia University Press）出版。

图20：
通过小桥、大台阶和柱廊的进入方式

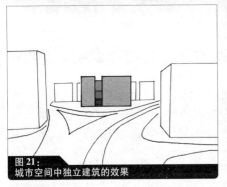

图21：
城市空间中独立建筑的效果

图22：
穿过半公共、半私密居住庭院的进入方式

进入场地、接近建筑

通常，你已经知道人们进入建筑场地的位置和方式。场地要么坐落于街道上，要么用通道与街道连接。功能需求的独立性、连接通道也决定了使用者和参观者感知建筑的方式（见图20）。人们走进建筑时会产生什么样的效果？这是一个需要考虑的重要问题（见"设计与文脉"之"社会与文化要素"）。

重要的一点是建筑与街道相联系的入口层所处的高度。如果入口层处在街道层的下面，那么连接入口与街道的通道看上去就不是那么重要。如果入口层比街道层高一些，则能让建筑显得高大、让人产生敬畏。另一方面，如果建筑是通过前院进入的，则入口空间会产生距离感，即使可能给人的印象更为深刻。通过院落进入的入口空间作为与公共空间的过渡，在建筑的前部形成了一个公共的领域（见图21、图22）。你还应该考虑残疾人的需要，尤其是在公共建筑中，要把入口布置在建筑物的首层。

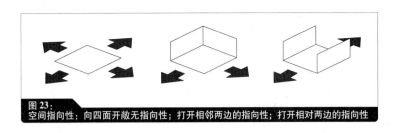

图 23：
空间指向性：向四面开敞无指向性；打开相邻两边的指向性；打开相对两边的指向性

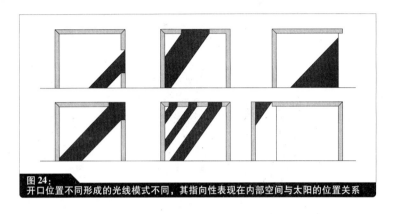

图 24：
开口位置不同形成的光线模式不同，其指向性表现在内部空间与太阳的位置关系

朝向

除了建筑的三维形体、在场地中的位置、入口的设置等，建筑的指向性是另一个需要考虑的方面。一幢建筑可以是封闭的、实体的效果，也可以是通透的、开敞的效果（见图23）。但是，如果这些不同的可能性缺乏指向性，那么它们各个面的表情就都是相同的。

根据不同的场地条件，你可以不同的方式处理建筑的各个立面。例如，你可以希望建筑指向太阳，或者根据不同的光照条件要求设计各个房间（见图24，见"设计与文脉"中"景观与气候"部分）。在温和气候地区，居室朝阳是人们所追求的，但是对于艺术家工作室或者博物馆建筑，朝向太阳就会成为缺点，因为这类空间需要均匀的、漫射的北向光线。

在建筑环境中还有其他影响建筑指向性的因素。如果建筑临街的一面比较喧闹，你就得对居住、休闲空间进行屏蔽和防护。建筑的后部可能有一个向公园开敞的视野，住户愿意从建筑内部去体验。或者让大量的公寓受益于景观的特殊特征，比如附近的河流（见图25）。这些特殊的情况都要求建筑具有特定的指向性，并且能激发出与场所相

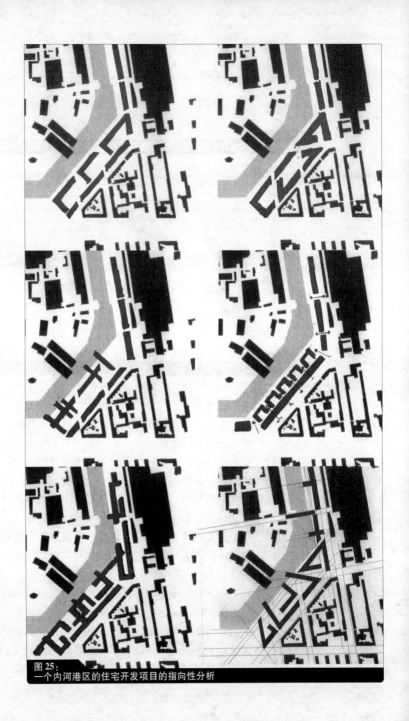

图 25：
一个内河港区的住宅开发项目的指向性分析

图 26：
外部空间的框景

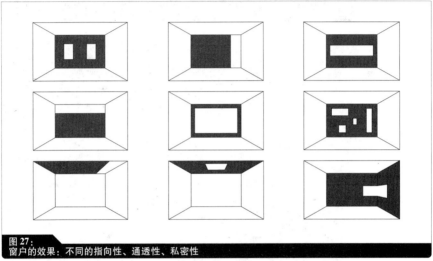

图 27：
窗户的效果：不同的指向性、通透性、私密性

关的设计途径。比如采用整片的玻璃墙来强调一个封闭房间的特殊景观，或者采用一系列小窗户来展现周围环境中某些特殊的细节（见图26、图27）。

室内外空间的关系可以形成不同的建筑设计原则，既可以是"封闭/内向的"，又可以是"开敞/内外交融的"。在密集的城市环境中，一个玻璃房子因为缺乏私密性因而是不宜提倡的。但把它建在一个广阔而空旷的风景中却不失为一个好主意——建筑最小限度地介入环境，与周围环境进行对话。

建筑与其环境的关系

如果一幢建筑想要融入周围环境，其方法不外乎是基于对建筑外部的感知，或者内外空间的交融。例如，全景窗可以用于加强山谷的

图 28：
将外部景观结合到教堂空间与功能中

图 29：
玻璃表面可以强调内外的交融（通过透明性），也可以表达封闭的效果（通过反射性）

美丽景色，或者通过精心摆布窗的位置可以把室外景观转化成室内要素。建筑也可以围绕着一棵大树来布局，这样大树就变成了室内空间的一部分（见"设计与文脉"之"风景与气候"）。

采用落地窗，或者为了区分空间的开敞与封闭而组织好观看建筑内外景物的视线，是创造空间体验的非常重要的设计工具（见图 28、图 29，见"设计与功能"之"满足使用者需求"）。

从建筑内部考虑将建筑立面设计得各不相同在创作理念中扮演着

图30：
室内空间和室外空间的自如转换

图31：
带有景观而不受视线干扰的外部空间

重要角色。如果外部区域有视线干扰，随之而来的私密性的缺乏会很快抵消其吸引力。然而，从被遮挡的区域看一处美景则会戏剧性地改变空间印象。同样可以通过设计，从视觉上隐藏室内外空间的界限，从而将私密性的外部空间结合到建筑中（见图30、图31）。

出于这个原因，建筑永远不应被当成孤立的事物考虑。如果可能的话，你应当对用地、建筑及其使用者之间的相互作用关系作一些研究。设计受着各种与场所有关的要素的影响，如相邻建筑的投影、每天从早到晚（每年不同季节）太阳光线的变化，以及建筑内外特殊景物的取舍。这些不同的方面同时也表明，设计往往产生于特定场所与使用需求之间的相互影响。

社会与文化因素

前面的章节论述了影响人们感知建筑的诸多方面。由此看来，将建筑被感官认知的方式与这些印象在人脑中被加工的方式加以区分是非常重要的。即使感官认知总是相同的，信息处理的方式也会因人而异，因每个人的个人经历和社会文化背景的影响而不同。这就解释了人们感受同一个场景或者建筑的方式为何是不同的。总体上说，知觉并不是客观的，即使在综合了大多数人的意见后会出现"客观化的主观性"。由于人们在很大程度上被社会化了，而且会受到各自教育背景的影响，因此这种主观性大都被局限在狭窄的社会、历史和文化的框架内。对于一个设计任务，从历史、社会、文化背景入手进行研究，通常也会产生出重要的洞见，可以为创意的生成提供灵感的源泉。

历史因素　　对于场所的研究不要局限在对场地本身和邻近地区的考察。每一

图32：
加建的玻璃大厅使古老的城堡遗迹获得了新生

图33：
新的文化设施与罗马神庙之间的对话

个建筑项目都是对场所历史的回应、对场所未来的规划。对现状的设计和改变是一种介入（intervention），这种介入不可避免地会被当作环境中持续演变过程的一部分。建筑物都是为某种特定目的而建造和使用的。它们可能被再利用、被改变、被扩建、被拆毁、被重建。某些时候，它们也可能保持闲置状态或者陷入荒废境地（见图32）。这些建筑通常具有重要的社会性地位，无论其是否曾被载入史册，或者某个人物与某个特定场所之间有关联。创造一个场所的历史参照坐标可以为设计概念的发展带来多种途径。你可以参照集体记忆（collective memories）——由社会整体记忆所形成的印象——也可以参照委托人或者先前的住户的完全个人化的故事（见图33）。同样，你还可以把你本人的经验结合到初始创意中，从而获得进入主题的个人化途径。

重要的是要知道，参照系反映着建筑设计任务的重要性。例如，如果一个设计任务具有社会意义——如博物馆、纪念碑等——就必须要关注与历史事件的联系。但是，如果设计对象是居住建筑或者购物中心，你就需要掂量对历史方面的考虑是否适于这样的情况。

社会文化背景

如果你的设计概念回应了社会发展，无论从总体的社会背景还是从特定的现象中，你都能够找到基本的方法。例如，当你设计一个住宅综合体时，你可以通过设置一条穿越场地的公共通道，或者通过把公共设施结合到项目中，尝试去回应公共空间私有化的问题（见图34）。进一步说，建筑的无障碍设计无论对城市发展还是对建筑的进入方式都是有影响的。

社会文化分析的一个有意思的方面在于，它给了你一个表达自己观点和思想的机会：老年社区需要混合什么样的功能？不同种族背景

图34：
居民对公共场所的使用

图35：
地中海地区的地域性建筑

的人群如何整合到社会中？我们的生活中社区意识和个人意识应该各占多大比例？环境保护和对气候变化的关注应该扮演什么样的角色？其重要性是否能够在设计理念中有所表现？如果我们希望避免两败俱伤的后果、希望避免景观恶化的失控，那么开发强度是否应该有个限度？

无论你选择什么样的方法，重要的是你要把最初的建筑尺度与广泛的文脉及价值体系联系起来。这将有助于你建立自己的建筑观。

地域性建筑

在历史的长河中，在不同的气候、文化、社会条件下演化出各种各样的地域性建筑形式和建筑类型，这种演化常常是由于材料短缺、特殊使用功能或功利性的形式造成的（见图35，见"设计与文脉"之"景观与气候"）。

这些建筑类型常常会针对各种需求提出出乎意料的简单解决方案。通过研究和分析地域性建筑，对于认识建筑与社会文脉之间错综复杂的关系，可以不断地得出令人惊讶的新洞见。非常重要的是，不要把建筑从它的功能、环境和建造的时代中孤立出来看待。

国际化

总之,我们不能拿当代的状况与那些许多传统建筑形式演化的时代相提并论。我们有更多的手段与材料来应对挑战,即使我们面对更困难的气候条件。使用功能和需求也有大幅度的变化。作为正在发生的国际化进程的结果,与过去相比,各地区、各城市之间的差异性如今很少被提及。建筑材料和建筑形式不仅变得越来越相似,而且我们的行为习惯也变得越来越相像,因为我们对其他国家、其他文化了解得越来越多。

为了发展可持续建筑,使之能与周围环境进行生动对话,你应该对地域的、本土的建筑的差异性有所了解并作出反应。传统的建筑形式和类型在这一过程中起主要作用,因为它们强烈地影响着新建筑在其环境中被观看的方式。例如,由于当地的建筑传统,一幢新的砖结构建筑建在德国的北部或者荷兰会很容易与周围环境相协调,而同样一幢建筑如果建在意大利南部的小镇则会格外突出,因为当地的建筑都是同样颜色的抹灰立面。地中海地区以温暖和对比丰富著称的特有的光线模式,使得具有三维形状立面的简单建筑显得更有动感。同样的建筑,如果放到莫斯科那样弱对比光环境下去看,则会完全不同。当代建筑的形式不是能够随便被交换或者被复制的,它们更加紧密地和场所交织在一起,而不像初看上去那么简单。

象征主义与图像化

建筑传达着信息。这种信息可以是显而易见的,能够被任何人辨识,也可以隐含起来逐渐地透露给观者。在建筑历史的各个阶段,建筑如何传达信息、信息又是如何与建筑的使用目的相关联等问题已经被人们探讨过。

这里有两种相反的做法:一所学校的设计遵循着忠实结构的原则,努力追求功能与形态的一致性。而另一所强调建筑在承载意义、创造个性方面所起的作用,不在乎使用目的和功能。这所学校使用符号化的、象征的个性化建筑母题,把这一概念发挥得恰到好处。

在折中主义和后现代主义建筑中,对历史建筑的参照并以此作为母题的做法是为了在观者中引起共鸣,以取得特殊的效果。例如,一幢建筑如果掺入了希腊神庙的形式语汇,或者以现代建筑形式为样板,那么它看上去就会更引人注目。这个设计就会使观者能够猜测出该建筑的使用目的、功能是什么?使用者是谁?这些猜测的正确与否并不重要。外壳和功能并不能形成一个创造性的整体,而且,以这种方式设计的建筑师总是设法把外壳分离出来,赋予它独立的量度,使

图36:
将结构和其外壳分离的原理

图37:
以居住建筑说明将外壳与其结构相分离的原则

之承载意义的作用得到强化（见图36、图37）。

与此同时，建筑师也可以采用那些与正式的参照物相平行的象征性的主题。例如在文艺复兴时代，曾经盛行于古代的人性的观念和价值得到回归。在建筑和艺术领域，这种回归表现在对古代建筑、古代原理与主题的阐释之中。尽管如此，必须要注意的是，一个信息从某个人的或者专业的眼光来看是逻辑的，但在其他人看来则可能完全是不同的。因此，在采用象征性、符号化的风格之前，你必须要分析其文化的、社会的文脉关系（见"设计的构成"）。

提示：
符号承载着意义，符号的价值取决于其内容、意义与形式的统一性。符号代表了一个对象而保留了其内在与外在的统一性。与此相反，当建筑师"引用"某正母题时，他们并没有打算去追求内容与形式的统一。

注释：
建筑的象征主义是个非常引人入胜而又错综复杂的题目。深入研究可参阅《向拉斯韦加斯学习》（Learning from Las Vegas），罗伯特·文丘里（Robert Venturi）与丹尼斯·斯科特·布朗（Denise Scott Brown）著，麻省理工学院出版社（MIT Press）1977年出版。

与其他学科的关系　　建筑与其他艺术形式之间的关系和相互作用可以通过研究各个时代的艺术与建筑史来得到更好的理解，要记住艺术与建筑在任何时代并非总是采用相同的原则和方法。其中一个最重要的差别是，除了它的艺术冲动，建筑必须满足人的基本的需求，为人提供能够遮风挡雨的生活、工作、休息的空间。即使建筑由于具有这些特殊功能而获得了比其他艺术形式更特殊的地位，建筑与其他艺术形式之间的关联性仍然可以作为设计概念的基础：

在艺术与建筑紧密关联的时代，我们能清楚地看到，一种原则可以从一个领域转换到另一个。在文艺复兴时期，绘画中因透视而被缩短的空间与实在的建筑空间之间有许多相似之处。包豪斯把建筑作为艺术的集成，其目标是把建筑与其他艺术形式结合起来，把所有艺术形式统一起来。后现代主义的形式主题，包括象征性和符号化，在其建筑中是最清晰可辨的。在当代艺术中同样也存在大量与此类似的案例。

理查德·塞拉（Richard Serra）在他用锈钢板制成的可进入的空间雕塑作品中，通过展现材料老化过程探讨了材料的本性，也强化了材料的重量感——他的很多作品的重量都超过了100吨（见图38）。这些雕塑令人震撼的空间效果和气氛反映了公共空间所引起的令人不安的感觉。彼得·埃森曼（Peter Eisenman）在柏林大屠杀纪念碑的设计中也采用了类似的手法。

> 提示：
> 艺术通常分为四个不同形式：
> 视觉艺术，包括绘画、雕刻、建筑和应用艺术。
> 表演艺术，主要为表演（舞台、电影）和舞蹈。
> 音乐，大致划分为声乐和器乐。
> 文学，部分为叙事文、戏剧和诗歌。
> 视觉艺术作品通常是具有空间感的实物，本身或内在有视觉效果，而不需要解说。

> 提示：
> 就空间感和实体感而言，室外雕塑显示出很多与建筑相类似的特性。物体与其周边环境有趣的互动关系可以被用作训练对空间的感知。如果艺术作品是为某个特定场所而创造，这种类似性将格外明显。

图38：
理查德·塞拉的雕塑作品

尽管如此，大多数艺术作品不能仅仅通过观察来理解。阅读艺术家所写的文字以及相关评论文章和传记同样是重要的。大多数艺术家写下了大量的文字来记述其工作方法和创作动机。尝试用其他艺术领域借鉴工作方法和设计方法会是一种很有收获的体验。通过研究绘画、雕塑、摄影、音乐以及其他艺术形式的各种技巧，你可以找到建筑设计创作的灵感。

设计与功能

建筑的功能通常对于建筑设计以及其深化具有决定性的影响。功能可以为你建立一个供你遵循的总体性框架，也可以作为设计概念的出发点。许多建筑师的设计是通过平面图或者空间框架图来满足特定的功能需求的。利用自己的创造性技巧和经验，他们从功能中提取出某种超越功能的特定的表现样式和形式语言，将形式和功能统一起来。

功能作为设计的出发点

从现代主义时期到现在，把功能当作设计的基础已经很普遍了。

功能在建筑设计中的作用也变得格外重要。在设计创作中反映功能并且在视觉上表现功能成了建筑设计的基本目标。

以下部分所描述的步骤可以作为工具和方法，将有助于学生理解功能并把功能结合到自己的设计中。原则上说，这些步骤并不是要求进行一种工艺设计（functioning design），而是建议要利用这些工具，尤其是当对所设计的功能不是太熟悉时更是如此。"工具"这个词本身已经强调了它所能提供的帮助并非处在设计工作的中心地位。在功能图解中描述相互关系预示着你已经完全把握了功能的复杂性。理解功能的方法是去阅读那些提供了典型空间需求和功能联系的读物（见"附录"之"参考文献"）。另外，你可以分析那些功能特殊或者相似的建筑实例。通过研究几个实例的平面图就能发现个别部分在布局、尺寸和结构上的相似性和共性规律。这些都可以结合到你的设计中。

满足使用者需求

当你把注意力转向建筑功能的时候，通常还会涉及到需要这些功能的人，无论这些功能是居住、工作还是休闲。重要的是去分析使用者在建筑中是如何实行这些功能的？或者他们希望得到什么样的体验？例如，设计电影院时我们应当认识到，人们希望从这一建筑中得到一个特殊的体验。他们希望得到娱乐，甚至沉浸到另一个世界中，从日常的生活中得到片刻的解脱。你的设计可以把这些期望考虑进去。相反，你在设计办公空间时，应该想到办公人员在工作的时候希望集中精力，不被干扰。这就需要满足他们对光线、通风、隔声以及空间分隔的需求。

研究使用者需求有各种方法：一方面，你个人要对使用者有一定了解，知道你在为什么人在做设计。另一方面，设计任务也可能集中在某个特定的目标人群（如住在公寓里的老年人）而不是某个具体的人。

个别需求　　如果使用者是已知的，你可以去研究他（她）的个人兴趣和要求。例如，如果设计任务要求你为一位艺术家设计住宅或工作室，可以通过和艺术家进行交谈、观察他（她）的工作方法来了解他（她）的愿望和需求。也许该艺术家想要一个安静的、光线均匀的、比较封闭的工作空间，也许他（她）更愿意欣赏外部的风景或者大城市的喧嚣。

聚焦目标群体　　当你为一个特殊目标群体设计建筑时，你会面对完全不同的情况。在这种情况下，你必须首先了解这一群体的总体需求。在这里，为了了解更多目标群体的需求，去研究当地做法或者去分析相类似的实例所面临的问题都是有帮助的。这些问题随后也会在你的设计中碰到（见"设计概念的形成"）。

与使用者打交道

在大多数情况下，你可以通过与使用者直接接触、通过和不同类型的使用者交换意见来研究使用者需求。你应该抵御住诱惑，不要试图仅仅从外部去了解内部过程（如对于某些特殊的工厂或消防站）。当你遇到你不熟悉的功能时，这一点尤其重要。那些整天和某种建筑功能打交道的人对事物总是有不同看法，当你把功能流程转化为空间设计时，把他们的看法考虑进去是很重要的。与一个机构中各个层次的雇员或用户充分地交流看法将有助于你分析程序、结构以及既有建筑中存在的问题。尽管如此，从使用者那里收集来的信息在设计中的用途是有局限的，因为他们缺乏建筑师关于设计的背景知识。正因为如此，设计一个功能完好的建筑的理想方式，是将使用观念和详细规划放到一起来研究。

空间布局与内部组织

特定的工程项目通常对应着特定的空间需求。例如，一个家庭对于他们想要设计的房子会有个特定的想法，一个公司会根据其员工的数量对工作空间提出要求，一家博物馆会根据特殊的展品来确定其展览空间。这些需求会决定必要的空间和体积，会对项目的规模和尺度形成初步的概念。

空间与体积

对于那些标准化的功能，比如公寓和办公楼，你可以通过累加结构面积、辅助面积，根据一定的平面系数，算出需要的建筑面积，并乘以标准层的层高来大致估算一下建筑的体积。

> **实例：**
> 假如一个家庭大约需要 120m² 居住面积，要计算总建筑面积，需要加上 20%～25% 的结构和设备所占用的面积，用层高 3m 去乘总建筑面积得出大约 450m³ 的体积。假如你设计的是一幢两层的立方体建筑，这就转化成了 8m×8m×8m 的立方体。像这样的思维练习可以使你形成建筑规模的感觉。

像室内游泳池、博物馆和活动大厅这类建筑通常包含着各种各样

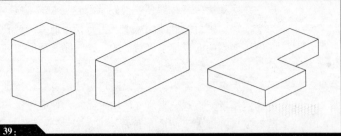

图39：
相同体积下的三种不同体量：塔楼体量、板楼体量、曲尺形低层带院落的体量

的房间，其中一些对于大小和高度有完全不同的要求。因此，在创作过程中，一定要分别考虑每一个部分以满足特殊的需求，设计出不同的特色。室内游泳池在一个建筑群体中可能是主体性的建筑，因此其辅助功能既可以组织到主体建筑中，也可以分开布置。

一旦你决定采用这种方式进行设计，重要的是，你不要把这种粗略的设计当作实际的设计。否则，你就会把粗略的形式当成已知条件，从而使设计过程陷入止步不前的境地。你一定不要忘记，建筑是个抽象的外形，其形态可以随心愿不断调整和修改（见图39）。

除了要大致估算建筑的空间和体积，还需要制作一个空间分配表，不但要明确每个空间的功能和需要的面积，还要对相关功能的空间进行分类（见表1）。

空间分配表示例　　　　　　　　　　表1

咨询中心

功能	房间面积（m^2）	数量	总面积（m^2）
接待区/助理	40	1	40
等候区	40	1	40
助理	12	2	24
主管	40	1	40
办公	20	1	40
团体接待	25	2	50
咨询室	25	17	425

续表

咨询中心

功能	房间面积（m²）	数量	总面积（m²）
会议室	30	1	30
职员休息室	20	1	20
厕所	40	1	40
二次销售面积		15%	109
总计		28	838

空间布局方案

对于许多设计任务而言，空间分配表已经是现成的了。即便如此，设计任务有时也可能要求对空间分配表按照实际功能需要进行调整。建筑师在实践中经常会遇到这样的情况，有些委托人对建筑的基本功能有明确概念，但他们并不清楚还需要哪些辅助空间和辅助功能。

因此，你要做的第一件事，就是要在确定主要功能空间的同时，确定所需要的辅助空间，如大堂、厕所、门厅、过厅等。如果你想对整个设计任务有个完整的考虑，你还要计算出每个区域所需的空间大小。

举例来说，如果一家公司想要为500名员工提供办公空间，你必须首先确定办公人员所需要的办公室类型（单间办公或者开敞办公）以及每个人所需要的空间面积。辅助功能和交通面积也必须同时考虑在内以便获得空间需求的总体概念。

在设计老年人住宅时，你可以采取类似的方法。你必须和委托人一道从一开始就要确定，是把个体住宅的功能组合成居住单元，还是按照所有住户的要求统一组织。

提示：
关于住宅功能的更多信息可参阅本套丛书的《设计与居住》（Basics：Design and Living），中国建筑工业出版社2010年出版（征订号：18807）。

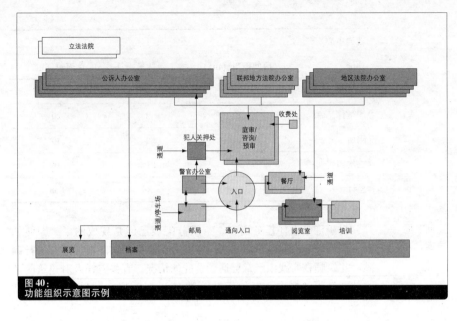

图40：
功能组织示意图示例

利用空间分配表，可以按照空间对高度、照度等方面的不同要求，对具有相同需求的功能或者空间进行分组。从该表能看出各类空间的分配比例，使你很快能对设计的规模和尺度有个概念。

内部流线
与采光

为了进一步确定设计方案，你要通过对交通系统的设计对建筑进行组织，为确定建筑的体积和形状提供依据。例如，如果一个办公建筑是由单个的办公室组成的，你可以通过研究自然光线能照射到室内什么范围来确定建筑的进深。于是这就成了确定建筑结构方案的基础。

注释：
空间分配表中的各个房间，应该按大小比例表示出来，这样就能使你在设计中对空间布局有个直观的感觉。

实例：
假设有一幢办公楼，房间布置在一条走廊的两侧，层高3m。建筑最大进深可以通过研究自然光线透过办公室落地窗所能照到的最大距离来确定（大约5.5m）。如果加上走廊的宽度（约1.8m）、外墙的厚度（约40cm）以及与走廊之间隔墙的厚度（每道约15cm），那么最后得到的建筑进深大约是14m。如果办公室布置成组团（combi-office）方式，那么建筑的进深就能达到16m。

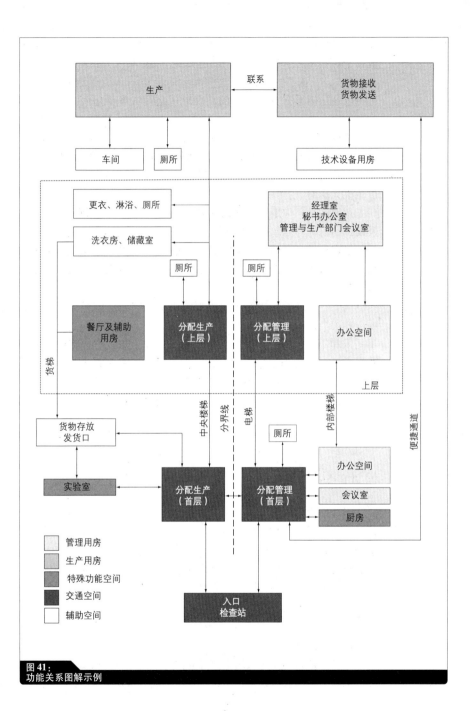

图 41：
功能关系图解示例

内部组织　　对设计任务中功能组织方面的考虑也是很有帮助的。为了确定建筑各个房间和各个分区之间的功能关系，功能组织关系需要用比较直观的方式表达。功能关系图解（functional diagram）是常见的工具，它可以通过图形来描述所有的功能分区之间或各个房间之间的关系（见图41）。

这种图解直观地表现了建筑内部组织关系，它表明功能分区和功能联系对建筑的形式是有影响的。比如，当私密的或敏感的功能分区要求必须与公共的或半公共的空间分开时，其空间布局必然会对设计产生很大影响。

建筑任务与目标　　重要的是，你不要把预先设定的空间布局方案与实际设计任务以及设计的目标等同起来。事实上，你应该按需要对这个预先设定的布局方案进行批判性的研究和修正。建筑任务书的内容比起纯粹的建筑功能要丰富得多，此外，上述方法还需要与其他决定因素联系起来看待。

例如，一座社区中心不仅需要一个带有附属空间的大厅，也需要一些必要的设备用房。为了满足功能需要，社区中心还必须提供不同人和不同团体会面、交往的场所。因此，除了交通组织和功能布局之外，软的因素也必须考虑。这些软的因素包括可达性、开放性、光线、气氛以及景观。这些因素经常会使设计任务与空间布局规划之间发生矛盾，而这种矛盾是可以通过改变空间布局规划来解决的。

P47
设计的构成

当处理形式与设计要素时，你应该牢记的是，一个设计对其他设计能够产生的影响是不可预料的。好的设计不仅仅由于满足了个别需求的所有方面，更重要的是由于它确定了个别元素之间的关系并赋予它们新的秩序。持续发展你的设计构思的惟一途径，是利用草图、图纸和模型去探索每一个可能的方案。

以下的章节集中探讨作为设计过程出发点的设计（见"设计基础"）。

P47
秩序与比例

从古代开始，比例就用来设计立面和整个建筑。历史上各个年代的建筑都对比例有所运用：从古代神庙的建造，到中世纪的教堂和文

> 注释：
> 建筑比例的历史发展是个令人着迷的主题。随着你对这个领域的了解，你会发现一些有趣的至今仍然有效的规律。如果你想研究建筑比例的历史发展，维特鲁威（Vitruvius）、阿尔伯蒂（Alberti）、帕拉第奥（Palladio）的著作都值得一读（见"附录"之"参考文献"）。

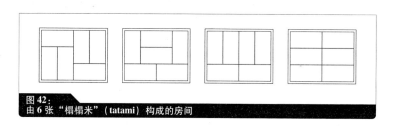

图 42：
由 6 张"榻榻米"（tatami）构成的房间

艺复兴时期的别墅；从经典的现代主义建筑，到当代建筑。

建筑师不断地利用数学规则去努力表现理想的比例关系。在古典时期，人们研究了神庙及其构件（例如各种柱式）的比例关系，特别是对那些与空间比率相关的方面很有研究。随后，对比例的研究继续深入。很多希腊的工匠和建筑师对几何知识非常熟悉，他们在建造神庙时利用数字的比率关系去处理个别建筑元素与其他元素的关系。

比例概念的双重意义早已包含在维特鲁威的定义之中（公元前 30 年）。一方面，"比例"这个词表示部分与部分之间的关系；另一方面，也表示和人体的关系。例如，一个多立克柱式下部的柱径尺寸与其高度的比例为 1:6，正好符合男子的人体比例。而一个爱奥尼柱式，其柱径与柱高的比例是 1:8，也对应着女子的人体比例。

日本"榻榻米" 　日本的榻榻米席子为建筑比例原理的应用提供了简单而有效的示例。榻榻米席子是日本传统住宅建造的基础。这种席子的尺寸一般是 85cm×170cm，尽管各地可能有些差异。由于每张席子都是 1:2 的比例，席子的排布方式可以是无穷无尽的。这些席子同时决定了房间的尺寸和比例（典型的日本和室大多由 6 张席子组成，见图 42）。

除了榻榻米席子的比例外，在建筑和许多其他领域还有比例关系可用（见图 43）。

黄金分割 　黄金分割率是最著名的比例关系之一。它表述为两个数字或者两

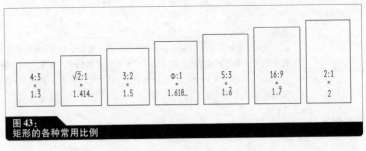

图43：
矩形的各种常用比例

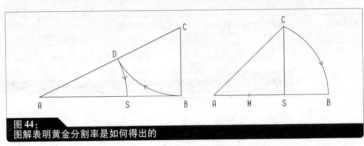

图44：
图解表明黄金分割率是如何得出的

个线段之间的比率大约等于1:1.618。和数学符号 π 一样，黄金分割率是个无理数，因为它不能用两个整数表示为一个分数：

$$a/b = (a+b)/a$$

黄金分割率还可以用圆来表示（见图44）。

模度

米制系统的引入终结了各种测量单位的统治地位。与古老的长度单位（如英寸、英尺、码等）不同，米制单位和地球的圆周相关而不是和人体尺度相关。结果，模数（modulus）和人体之间的直接关系作为比例的两个含义之一不再适用。为了把人体尺度重新引入建筑学，勒·柯布西耶根据人体的平均比例和黄金分割率提出了他自己的度量系统——模度（Modulor）。模度使得设计的比例和尺度与使用直接相关联。例如，桌面与栏杆的高度、窗的比例、整个房间和立面等）（见图45）。

体验比例

除了数学方法和分析，比例为观者创造了一种主观感受上的愉悦。尽管如此，每个人都希望获得系统化的、可以直接采用的法则，大量看起来设计得不错的建筑立面并非基于任何可以理解的数学法则。因此，当建筑师在确定各种建筑元素的尺度和构图以及确定二者的关系时，要取得均衡的比例也总是意味着他自己首先得有个良好的感觉。

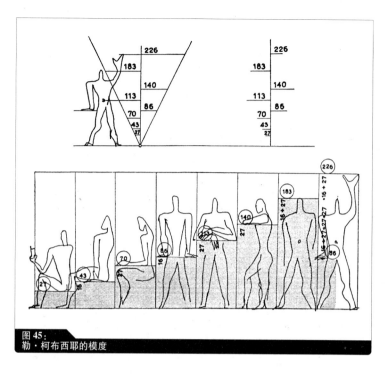

图45：
勒·柯布西耶的模度

例如，当你确定立面比例的时候你会发现，你可以求助于大量熟悉的规则和比例关系。虽然如此，在很多情形下，只有当你开始应用某种规则或比例，或者不断尝试直到获得满意结果的时候，立面才可能获得良好的比例关系。这可能包含对一面全景大窗进行移动或者改变，直到你对室内空间和立面的比例感到满意、并且最终与整个设计相协调为止。

提示：
如果你想知道更多关于模度的发展以及模度测量系统的背景知识，可参见勒·柯布西耶著《模度》（Modulor）第1、2卷，中文版将由中国建筑工业出版社于2010年出版。

注释：
你可以使用卡纸、橡皮泥、木头和泡沫塑料块制作简单的工作模型，去表现你想要的形态和比例。关于工作模型和模型制作材料，更多信息可参阅本套系列教材中的《模型制作》（Basics：Modelbuilding），中国建筑工业出版社2010年出版（征订号：18843）。

P50
几何形态

设计的基本元素

学校里所教的几何学是处理点、先、面、距离、角度等问题的。通过由欧几里德（Euclid，约公元前 365~300 年）建立并由后人发展完善的公理，你可以得到作为你的设计基础的形式。例如，二维形式（面）包括三角形、方形、矩形、圆形和菱形；而三维形式（体）包括的形式有立方体、直角平行六面体、球体和锥体。你可以运用这些基本的数学形状，通过变形、加法或者减法设计出各种各样的形式。

几何平面

物体表面的几何属性很大程度上可以搬用到建筑的表面。一个带有四个同样长度的立面的方形平面，非常适合设计成一个没有方向性的建筑，如公园里的亭子，或者空地或广场上的孤立建筑。圆形建筑更是如此，它不仅没有方向性，而且还指向自身的中心，从而提高了空间的重要性。相反，矩形的建筑有其自身的指向性。它的单向性使得立面有了正、侧之分。侧立面形成了特殊的三维的边缘，而正立面则形成了终止的尽端。另一方面，椭圆形建筑形成的空间有两个聚焦点。和圆形建筑一样，它也只有一个连续的立面。然而，和圆形相反，椭圆（和矩形建筑没什么两样）有指向性。如果你基于这些几何形状设计平面，你就能够了解这些法则，并且有意识地应用它们。这里所描述的原理可以在图 46 的实例中看到。

几何体量

在上述二维平面上增加一个坐标轴，你就可以创造出各种三维的实体。几何体和表面所遵循的规则是相同的。

简单几何体是非常令人震撼、个性鲜明的，它们尤其适合用来设计空旷环境中的"单体建筑"（object-like building）（见图47、图48）。如果把它们放到密集、混杂的城市环境中，和其他建筑靠得很近时，就不会那么容易了，效果往往也不甚明晰。

提示：
　　数学自从古典时期以来一直在建筑中扮演着重要的角色。达·芬奇所画的人体比例图（1492 年）在人体和几何形态之间建立了密切的关系。在古代，建筑师就利用基本几何形态设计总平面、平面、立面以及整个建筑。

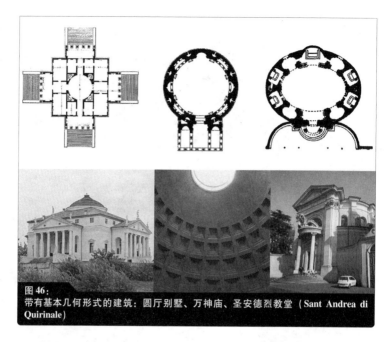

图46：
带有基本几何形式的建筑：圆厅别墅、万神庙、圣安德烈教堂（Sant Andrea di Quirinale）

图47：
直角立方体在环境中看上去很简洁

自然界中的形态

除了那些人类建立的物质的和非物质的关系，某些形式和结构可以得自于自然界，并且给你的设计带来一些创意。自然界中的结构通常都是环境中的有机形状或形态所构成的。这些结构通常都带有迷人的、变化丰富的轮廓线。大多数自然要素——即使它们第一眼看上去有些无序和随意——都遵循着某种清晰而又复杂的模式和规则。假如你观察它们的细胞构成，你会发现其结构非常独特，在弹性、承载能力、节约材料以及雕塑性外观方面都很出色。如果你能够从自然界获得一些创意，你同样也将很快在大自然的形式语言以及多样性方面找到一些启示。

图48：
采用几何形态的示例：圆柱体与锥体

图49：
新艺术风格的形态受到了自然界的启发

新艺术运动　　19世纪末、20世纪初兴起的新艺术风格（Art Nouveau）建筑，表现了其设计者对大自然的兴趣。各种各样的新艺术风格建筑展示了建筑师在自然界中获得的各种灵感：从植物母题装饰、动感的平面立面设计，到整个建筑的有机设计（见图49）。

此外，如果你有意把空间组织与结构原理转变成实在的建筑，研究大自然是非常有益处的。

> 提示：
> "仿生学"（bionics）这个词是由生物学（biology）的第一个音节和技术（technics）的第二个音节组合而成的。仿生学致力于把自然界的方法、结构和发展原理应用到技术系统之中。

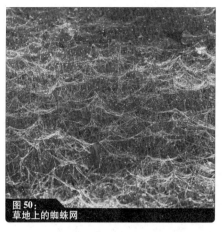

图50：
草地上的蜘蛛网

图51：
以索网结构建造的屋顶

仿生学　　　　许多建筑师利用仿生学原理来发展和优化其建筑与结构设计。费雷·奥托（Frei Otto）在慕尼黑奥林匹克公园设计的帐篷状索网结构、格里姆肖（Grimshaw）为"伊甸园计划"（Eden Project）设计的气泡状短线穹窿结构以及卡拉特拉瓦（Calatrava）设计的那些桥梁和建筑（其承重结构看起来像是骨骼），都是从自然界存在的结构原理中发展出来的。如果你要把自然转化为技术，就必须不断地对其进行修正和抽象化，因为自然形态实在是太复杂了，不可能被毫厘不差地复制。对自然形体背后的原理进行观察和分析是非常重要的，只有这样，从中得出的灵感才能被转化成建筑。费雷·奥托就是通过对钢索和肥皂泡的实验得到的屋顶形式。他所创造的形式完全是利用重力原理，完美地再现了力在屋面中的传递路径（见图50，图51。见"设计中材料与结构的应用"之"材料和结构作为设计元素"）。

　　　　动、植物展现出来的千姿百态的特征多能够被应用到建筑领域。如动物的骨骼、昆虫的复眼、犰狳（armadillo，一种杂食性的，掘河隐居的贫齿目哺乳动物，属犰狳科，生于北美洲南部和南美洲，特征为全身有连续的角质鳞片组成的盔甲状保护层。——译者注）的鳞甲、鸟类的翅膀等（见图52～图55）。

自由形态　　　　如果你尝试通过对结构与构造的实验得到设计的形式，你就能够创造出那些看似无序继而又会被其新奇打动形式。在创造自由流动的形态方面，建筑学与其他艺术形式越来越接近。这种方法尤其适合大型的、带有重要社会意义的建筑项目，如教堂、博物馆、文化中心等。设计过程的目的常常是使建筑更具动感、使建筑内外的动态因素

125

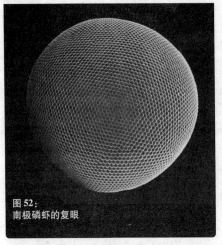

图 52:
南极磷虾的复眼

图 53:
位于英格兰康沃尔郡（Cornwall）的"伊甸园计划"

图 54:
天鹅展开双翼

图 55:
里昂的 TGV 高铁车站

图 56: 现代建筑中带动感的自由形态

变得清晰可见。这种策略常常用于那些倾斜的、带有圆角的建筑，或者那些带有雕塑性外观的建筑，或者那些带有连续性内部空间的建筑（见图56，见"空间与实体"以及"设计概念的形成"）。

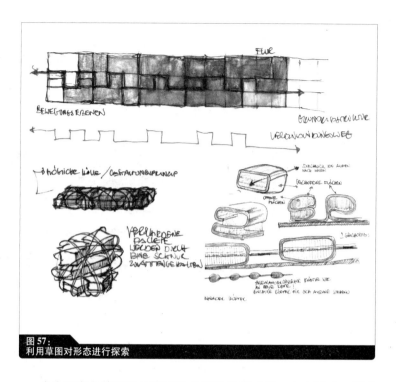

图57：
利用草图对形态进行探索

自由形态尤其对赋予建筑以象征性特征或使一个空间或一幢建筑具有独特性外观十分有效。然而，这种方法也存在某种风险，它会导致建筑师对建筑的形式过分关注而对建筑的其他需求视而不见（见图57）。

空间与实体

建筑始终是空间与实体相互作用的结果。实体限定了空间，物体只有在空间中才能被识别。空间和实体二者是相对立的，同时又是相互依赖的。只有空间与实体的相互作用才能构成一个整体。建筑也是二者相互作用的结果。空间与实体的相互作用所形成的空间次序、空隙、内部与外部，常常在明确限定的空间之间形成流动的空间序列。

假如你想设计一个复合形态，你会发现利用不同的要素有着各种各样的方法。你可以通过对几个元素进行排列组合来创造空间，你也可以通过对基本形态进行削减或变形来创造空间。各种手法的组合是无穷无尽的。

> 注释:
> 弗朗西斯·秦(Francis D. K. Ching, 一译: 程大锦)的著作《建筑——形式、空间与秩序》(Architecture-Form, Space and Order)在空间与实体的基本关系方面有很经典的论述(见"附录"之"参考文献")。

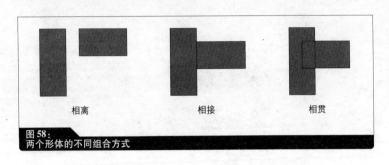

图58:
两个形体的不同组合方式

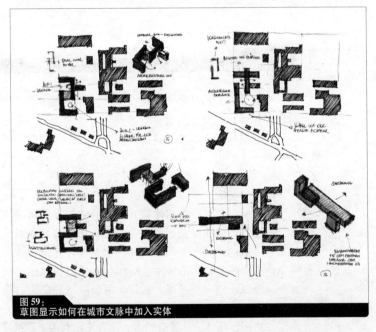

图59:
草图显示如何在城市文脉中加入实体

要素的分布与连接

假如你想通过"加法"来创造一个复杂形体,你必须要在你的设计要素之间建立关联性。你可以选择一些各不相同但互相作用的独

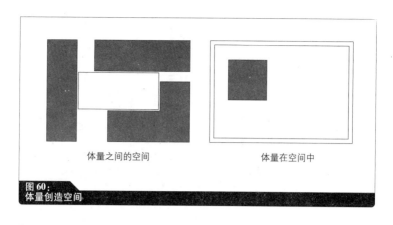

图60：
体量创造空间

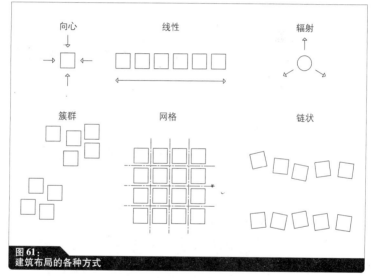

图61：
建筑布局的各种方式

立形态,或者一些同样的体量,以某种规则组织起来。

形态与形态之间可以相贯相切,可以相附相依,可以成行成列。它们总是相互作用并且创造空间结构（见图58,图59）。在这里,你有两种基本选择:要么通过布局在两个元素之间创造空间,要么在给定的空间中加入一个新元素（见图60）。

除了两三个形态的组合布局外,你会发现某些情况下（例如在城市或住宅开发中）更多的形态必须协调布置。典型的布局方式包括（见图61）：

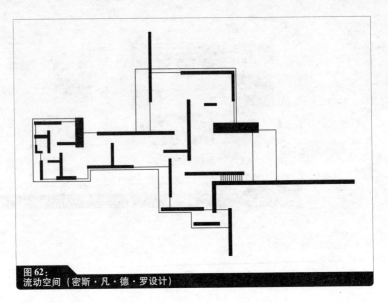

图 62：
流动空间（密斯·凡·德·罗设计）

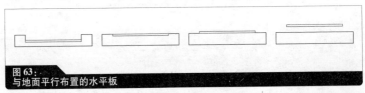

图 63：
与地面平行布置的水平板

图 64：
现代建筑名作中的平板

— 行列；
— 网格；
— 簇群或组团；
— 向心布置；
— 辐射状布局；
— 链状布局。

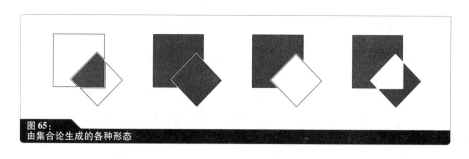

图 65：
由集合论生成的各种形态

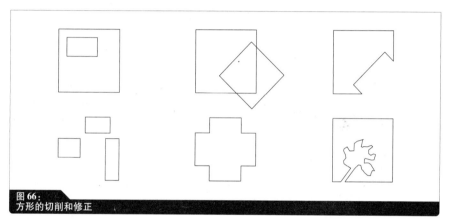

图 66：
方形的切削和修正

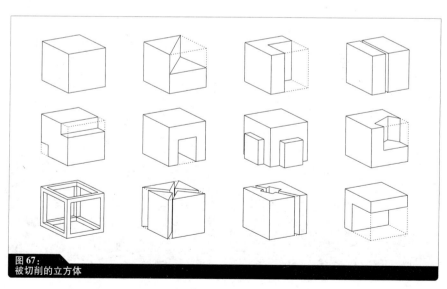

图 67：
被切削的立方体

用平板构成空间

平板（slab）是建筑中比较特别的形态。它可以水平布置也可以垂直布置。它也可以用来创造一个路径把人导向某个目标。根据排列布置方式的不同，平板可以用来联系室内外空间从而创造流动空间（见图62）。

你不但要考虑平板在平面上的布局，还要考虑水平板侧面给人的印象，因为这对设计有决定性的影响。基准面在竖向上的位置对于建筑的空间效果起着决定性的作用。它可以（见图63，图64）：

— 下沉至基础；

— 与地面齐平；

— 直接放置于地面；

— 悬浮于地面之上。

减少与修正

你可以通过体量的调整或切削形成表面和体量。如，一个零散的体量可以作为设计的几何学基础。方、圆等简单的基本形态也可以有无数种变化和改变的可能（见图65~图67）。

许多这类变化都是基于从集合论中得出的简单的数学原理。如果两个体量或者表面相交，就提供了许多可能性，创造了子集、交集、并集和差集。

折叠与弯曲

另一个改变或变换体量和表面的方法是折叠或弯曲。通过对一个条带的折叠和弯曲，你会得到一个引导性的空间流，使空间发生转

图68：
体量和表面的弯曲

图69：
折叠作为设计的原理

向。通过对体量的旋转和弯曲,你可以中和其几何硬度,从而使之变得柔软可塑。此外,你可以根据外部影响和作用力来改变基本形态,这些影响和作用力只有当形态被改变时才会被察觉(见图68,图69)。除了通过折叠和弯曲来改变几何元素的形态,你还可以利用计算机去生成和处理更为复杂的形体,自由地创造新的形式。

P63
设计中材料与结构的应用

从历史上看,设计的起点是结构,这一点在工业革命初期再一次得到验证。总体上说,建筑师确信,结构作为建筑的基本框架,在最终的设计中是可见的。在框架结构中,建筑可以采取柱网的形态来确定立面和室内的体系。自工业革命以来,工程结构如桥梁就已经公开展示其结构设计,使承重结构关系清晰可见。这一发展在19世纪末导致了一种新美学的出现并终结了掩盖、装饰结构的做法。这一点在建筑的许多领域都能看到。在这种情况下,在建筑结构与材料的使用上,存在着一种直接的和闭合的关系。例如,没有钢铁的工业化生产,带有细工装饰的工业建筑将是不可想像的。没有钢筋混凝土,壳体结构将永远不能建成现在的模样(见图70)。材料的技术的、视觉的和触觉的属性可以有意识地用于增强设计的表现力。

P63
材料和结构作为设计元素

结构可以用于设计的生成和组织。一方面,你可以根据文脉和功能形成一个设计,另一方面,你也可以先创造某种形式,然后再以某种不易察觉的方式去考虑结构上的需要。当建筑有某种特殊结构要求时(如大跨度),分别处理功能、形态与结构问题是非常困难的。因

图70:
钢筋混凝土双曲抛物面

图71：
分别由塑料、砌体、混凝土、木材建造的结构表现了不同材料的特性

此，我们建议将承重结构当作设计过程以及所有深入设计的起点。

如果你想把结构当作设计的基础，就必须对承重结构构造的技术标准有所了解，即静力学体系、材料的选择以及构件的尺度。

结构与材料的选择

总体来说，用于建造的材料是在设计过程中被确定的。然而，材料可以作为设计的基础。天然的石材可以给予周围环境一种特殊的品质，或者周围建筑由于气候原因选择用木材建造。（见"设计与文脉"之"社会与文化因素"）材料的选择也许受到你所要设计的初始构思的启发。如果你的设计以某种特殊材料为基础，在方案发展的过程中，你就要对材料的特性给予重视（见图71）。

结构作为设计要素

总之，结构不但建立秩序，承受荷载，在许多场合下，也能成为设计的核心特征。设计应该去强化结构的承载能力和材料的使用。设计也可以对美学表现以及建筑作品的结构产生决定性影响。

一旦你了解了需求和静力学条件，你就能开始认真考虑工作中真正有创意的部分。在创造性地利用承重结构的方面有许多不同的方法。例如，跨距可以被桁架、井字梁、双向板、悬索或壳体所覆盖。这里的每一种承重体系都决定着设计的决策和建筑的空间效果。即使空间上的考虑需要特定的结构承重体系，你仍然可以有目的地应用这些体系去发展超越基本功能的形态和结构。你也可以特意为某种材料的使用去设计各种不同类型的连接体，如铰接支撑和直立节点等。

如果你要设计一个不依赖标准构件的有趣的建筑，需要对静力学有足够的了解。然而，这里最重要的并非是要你能够去进行精确的结构计算，也不是要你去掌握结构和静力学原理。通过重力实验，你同样能够提出有创意的构思。在小模型的帮助下，不需要大量的结构计算，你能够发展出内在稳定的静力学系统。

提示：

有关材料和结构选择的指导内容可参见本系列教材中的：

《建筑材料》（征订号：18810），

《承重结构》（征订号：18857），

《木结构施工》（中国建筑工业出版社 2010 年出版），

《砌体结构》（征订号：18859）。

注释：

西班牙建筑师圣地亚哥·卡拉特拉瓦为他很多大胆的建筑制作了静力模型，以作为研究基础。如果你对这种发展设计的方式感兴趣，可以从卡拉特拉瓦的静力实验中获取不少灵感，来发展自己的创意。

结构反映材料

通过研究某种材料的结构规则和材料特性，你可以获得一些设计和形式的参数。例如砌体结构在连接较大跨距方面是有限的，所以经常以拱券的形式来使用。另外，如果你要建造一个木结构建筑，你就必须考虑材料的同轴性，并采取结构措施防止木材变形。钢结构允许跨度大、用材省，然而其防火性能却差强人意。不论你选择什么样的材料，建筑必须要考虑材料的特殊品质，并把它发挥到最大程度（见图72）。

模数网格与跨距

一旦你选定了材料，材料的表现力与功能需求则决定了设计中的跨距以及各种不同结构元素之间的距离。柱网结构是由可以重复计算的面积决定的（如果确实需要而且能形成设计的基础）模数化的柱网通常用于系统的组织、立面与空间。

柱网可以通过轴线的直角相交形成矩形区域，你也可以考虑使用任意角度的以及菱形的柱网，或者等边三角形的柱网。线性承重结构柱网的形成是基于一种线性的、没有第二个向度轴线的空间（见图73）。

在这种情况下，立面及非承重墙是居于轴线之上还是偏离轴线就成了问题。如果结构形成了一个框架和实体结构，墙体和支承体都具有承重功能，那么，墙体就应该布置在轴线之上，墙体就会被打断而

图72：
承重结构最大限度地发挥了其建造材料的特性

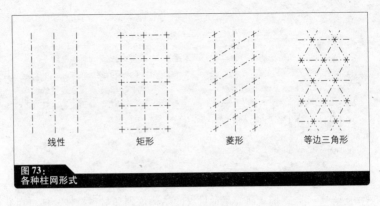

图73：
各种柱网形式

线性　　　矩形　　　菱形　　　等边三角形

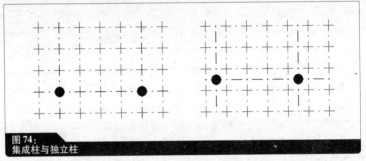

图74：
集成柱与独立柱

形成柱列。在纯框架结构中，通过墙体和立面从轴线的偏移，你可以清楚地区分哪些是承重构件，哪些是空间围合构件。这种带有一些独立柱子的结构具有更强的表现力和自治性（见图74）。

对材料的感知

P67

材料的不同，不仅是取决于其技术特征，还取决于观者的感受。我们对材料的感知是通过各种感觉器官的交互作用，方式是各种各样的。

直观感受　　由于人类近90%的外部刺激来自视觉，因此建筑及材料的视觉效果在建筑历史中一直扮演着重要的角色，并且一直被当作研究的主题。

材料与氛围　　使用常规的设计方法和工具去把握和呈现一座建筑在听觉、嗅觉和触觉方面的品质是很难的。这些品质从建筑的功能与文脉中是很难被感知到的，因为它们与建筑的材料特性密切相关（见图75）。

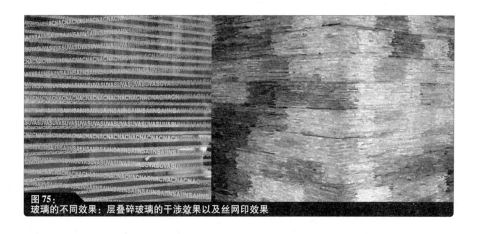

图75：
玻璃的不同效果：层叠碎玻璃的干涉效果以及丝网印效果

图76：
虚幻、飘浮的玻璃体

图77：
轻质结构和石材使新与旧形成对比

  对材料的娴熟掌握和正确使用可以支撑你的设计理念，甚至可以使你的设计理念得以实现。这些重要的材料属性包括：
  — 坚实的、朴实的；
  — 轻盈的、飘浮的；
  — 隐秘的；
  — 多层次的；
  — 透明的、半透明的；
  — 开敞的、封闭的。
  使用玻璃，你可以或多或少地消解一个轻盈的体量。而使用土质的材料，你可以强化一个由减法形成的建筑形态。如果你使用灵活的、透明的建筑材料，你就能创造出一个建筑外壳与建筑本身分离的印象（见图76，图77）。

设计概念的形成

总体来说，在实践中形成一个创意有多种途径。设计过程的开始常常是无意识的，或许某种经历或事件启发了你的兴趣、激发了你的想像力。那些"事件"一般来说就是前述的各种各样的外部刺激。这些外部刺激影响着建筑师的设计创作，也构成了建筑师可以吸取的经验基础。一旦你有了一个初步的想法，你就需要考虑如何去执行它，并且要从各种不同的途径去探索其潜在可能性。

忠实于理想

在设计过程中，你常常会发现你不能从始至终地、不偏不倚地坚持最初的构思。各种不同的外部条件以及基本概念需要尝试多种多样的设计方法和途径。你可能以不同的方法得到富有启发性的想法，并且希望能继续深化。当然，当你试图把一个好的理念与其他想法相合并，或者面对这个好理念犹豫不决时，这个好的理念就有被冲淡的危险。你应当时刻牢记你的初始想法，即使这样有可能牺牲其他的想法。为了另一个想法而放弃一个好想法并非毫无益处，因为简明易读可以强化你的设计品质。当你开始着手一个新的项目时，你会发现又回到了先前曾经放弃了的想法上。你的构思以这种方式拓展了你的知识。

简单而不乏味

坚持一个想法、寻求简明的理念也需要承担平庸乏味的风险。为了追求设计明晰性，往往会牺牲其复杂性和微妙性。正如建筑的功能需求一样，设计与建筑必须是多维的、多样的。然而，多样性也要求在细部的层面表述其完整性。许多有意思的设计都是从一个基本概念出发一直贯彻到最后的细节上。也就是说，你的基本想法不要变成一个教条，并不是说为了一个惟一的原则就必须要去除所有其他的。要确信你设计中的限定因素对设计构思要进行连接而不是进行转移。不要忘记，不同的原则在设计的不同部分可以相互重叠，相互之间起着补充和加强的作用。

表现的工具和技巧

你所选择的表现技法和工具可以对设计过程构成极大的影响。在制作一个物件时，最后的工作结果取决于这些工具。

铅笔是用于画平面、总图、结构框架、地形等高线的非常好的工

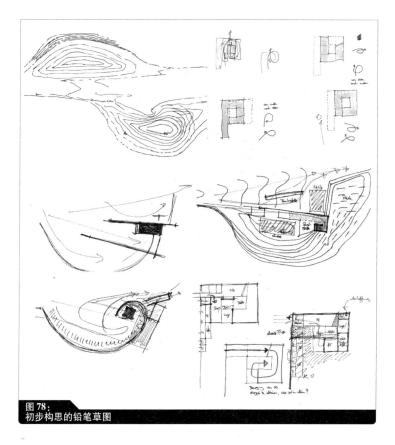

**图78：
初步构思的铅笔草图**

具（见图78）。平面图为研究空间结构、水平流线以及设计的功能组织和分区提供了一个理想的平台。剖面图为竖向关系和空间尺度的推敲提供了机会，而立面图为研究房间的日照、建筑内部与外部的关系、建筑的外观形象提供了可能。此外，假如你采用了新颖而奇特的媒介与表现技法，你就会得到一些新奇而有创意的解决方案。（见"设计概念的形成"之"方法与策略"）过去10年以来，计算机辅助设计（CAD）使得以前用铅笔等工具难以表达的建筑形式的创造成为可能。一个特别的新趋势——其本身也是数码世界多重可能性的一种产物——被称为"团块建筑"（blob architecture），它采用自由曲线（spline，多义线）形成复杂的、流动的、通常是圆乎乎的、生物形态的体量。只有采用现代的设计与视觉化软件才能使这类设计成为可能（见图79）。

图79：
计算机生成的形态（blobs，团块体）

建议建筑学生去尝试各种方法，以便发现哪种更适于直觉的方法，哪种更有利于激发他们的创意。

P72　　　**创造性与创新技巧**

当学生初次开始学习时，他们通常会怀疑自己是否符合学校对他们的要求，是否具备一个设计师应有的创造性素质。在设计课堂上，他们试图通过以前的经验来判断自己的创造性和设计能力。对于建筑设计而言，多大程度的原创性是必需的？创造性在设计过程中究竟扮演什么角色？

注释：

　　抛开CAD程序的潜力不论，许多建筑师发现用该软件很难进行设计，因为它们不能用于直觉（与铅笔和其他工具相比而言）。初学者通常都有这样的体会。由于这个原因，建议只有当设计的手工部分（如使用铅笔）不再吸引你的注意力，或者当你找到一个能用直觉设计体量与物体时再使用CAD软件。

提示：

　　如果你对"团块建筑"的相关信息感兴趣，我们推荐你阅读杰弗里·吉普尼斯（Jeffrey Kipnis）的文章《走向新建筑》（Towards a New Architecture）以及格瑞格·林恩（Greg Lyn）的文章《曲面建筑：折叠、弯曲与柔韧》（Architectural Curvilinearity: Folded, the Pliant and the Supple）。两篇文章皆出自《当代建筑理论与宣言》（第二版）（Theories and Manifestoes of Contemporary Architecture）。

创造性   在设计的学习过程中,你不可能像站在一块空白的画布前,去"画出"任何你想要的东西。有一些边界条件影响着设计进程,并且决定着设计的出发点和发展方向。在设计的过程中,你很少会形成以前从来没有存在过的想法。真正的难点在于在已有的设计原则和方法的基础上形成新的解决方案。因此,不考虑个人潜能等因素,创造性往往来自于外部影响的激发,来自于那些已知的事物以及那些要继续探究的愿望。

    科学界已经发展出的各种创新技巧对建筑设计非常有用。这些技巧的目的在于在短时间内直觉地生成大量的想法,在于激发联想和针对问题的新的思想方法,还在于激活隐含的想法并将最大限度地减少对创造性的压抑。团队成员之间的互动是发现新解决方案的有效途径。

头脑风暴   在"头脑风暴"(brainstorming)会议中,要建立一个小组,并且要明确任务。首先,要对任务进行解释与分析。如果需要,也可以先提出一个典型的解决方案。然后,所有成员都要提出自己的见解和想法,并且要求不对他人的想法提出评论,也不要对他人的观点进行判断。只有这样,团队成员才能够不再一开始就去思考一个设计概念的所有结果,才能享受到一种自如表达自己想法的自由。即使是一个荒谬的想法也能激发其他成员产生另类的方法,进而鼓励他人去刺激更多的人。会议过程中所有这些想法都被记录在案,然后,根据可实施的程度进行展示和修改,并最终达成决议。

头脑写作   头脑写作(brainwriting)和头脑风暴具有相同的程序。区别在于团队成员的想法不是直接表述给团队,而是各自写在纸上。这样就使得那些不善言辞的成员能够更容易地表达他们的想法。

> 提示:
> 人们通常会把创造性这个词与用新颖的、有效的非常规的方式解决一个问题或者完成一个任务联系起来。没有创造性,某些问题很有可能永远无法被识别或被解决。进而言之,创造性有助于你更灵活地去接近一个问题,并且获得以前在某种条件下无法获得的新的资源。

展示法	还有一种常用的鼓励创新的技巧,叫做"展示法"(gallery method)。这种方法在建筑学的课程中尤其常用。每个参与者把想法写下来,然后张贴到一块展板上。这些想法随后在一个讨论会上进行讨论,接下来再根据讨论中出现的新洞见进行修改。最后,修改后的想法再次展示出来接受小组其他成员批评。
SCAMPER法	SCAMPER 是一种类似核对表(check list)的创新技巧。其目的是发现新的方向和质疑普遍性的想法(见表2)。通过对特定问题的回答,参与者可以尝试去发现一些设计途径,从而走出可能的误区。

SCAMPER 法　　　　　　　　　　　　　表2

缩写	含义	
S	替换(Substitute)	替换个别元素
C	结合(Combine)	与其他元素进行结合,或者结合各个不同元素
A	改变(Adapt)	对内容和功能进行改变
M	调整(Modify)	对尺寸或比例进行调整,使元素多样化
P	置入(Put)	置入其他使用功能
E	消除(Eliminate)	消除附加元素,只剩下核心功能
R	倒转(Reverse,原文误为Reserve——译者注)	倒置,直接反对该想法

思维映像	思维地图(mind map)是一种提出任务的图示。核心任务写在纸页的中央,并且要与从属的方面和变化有视觉联系。例如,假如你试图研究一个设计任务及其功能,思维地图可以帮助你提出功能配置图(见图80,见"设计与功能"之"空间布局与内部组织")。
单个设计师的工作方法	即使你不是以团队的方式工作而是独自探索设计概念,你也会发现上述创新技巧是有用的。为了避免在设计过程中被创造性束缚的危险,当设计成形时最好能够审视和质疑你自己和你的设计,像局外人那样以不同的眼光和角度看待这些不同的方面。和他人的讨论(包括其他领域的人)是很有好处的。你还可以把自己放在未来使用者的位置,尝试从他们的眼光来看待设计,从而了解建筑对他们的影响。

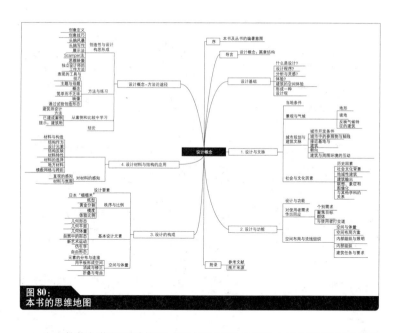

图80：
本书的思维地图

为了使你与你的设计构思或者与你所追寻的某个特定思维线索保持一定距离，建议你经常中断一下设计过程，让自己被其他事情占据片刻。这样就能使你以局外人的眼光挑战和批评你的设计中潜在的理念。

P74
分析与调查

方法与策略

通常你一开始要面临的任务是研究上述的设计边界条件。没有人会知道哪一条途径能够成功。最好的办法是去参观场地。当你画草图、拍照片、研究周边环境时，你会产生设计的第一个想法。通过分析空间与城市肌理以取得对一个地段的感觉是很值得尝试的。尤其是当你面对一个陌生的场地时，这种方法尤为重要。

模型与建成案例

另一种方法是去研究那些解决类似问题或处于类似环境中的建筑案例，以此来获得一种感觉。当设计任务中包含一个特殊的使用功能（如一个火车站，或者处于坡地上的建筑用地）使得对周边环境的考察很难进行时，对已有案例的研究则为你解决该问题提供有用的线索。你研究的建成案例越多，你的选择余地就会越大。当设计往前发展时，你要记住美学敏感性不是需要考虑的惟一标准。你还要考虑总体状况，尝试去显现和分析一座建筑在其特定文脉中的意义，从而发展出自己相应的概念和方法（见"设计与文脉"）。

图81：
"轮船"的母题在不同时代建筑中的应用

图82：
"进化"的母题被演绎成了盘旋坡道，博物馆观众可以沿着展览主题穿越建筑

图83："岛屿"的母题用于林中住宅项目的设计

著名设计师的设计方法

  还有一种方法是研究著名建筑师的作品，然后尝试通过应用这些著名建筑师的理念和构思来发展自己的方案。当你设计一个方案时，你会对这些建筑师的作品和观点有一个更深入的理解。通过在你自己的设计中应用这些内容，你就会与之形成某种关联，并且提出你认为适合自己的方法。

主题与母题

  建筑常常不会表现为一个可以确认的设计概念。这是由于建筑要承担大量的功能，而这些功能在设计过程中往往被同等地对待。为了使你的设计具有个性，你可能赋予它一个母题，或者去寻找一个好的基础平台。有的时候，母题可以在建筑设计之外的领域找到。

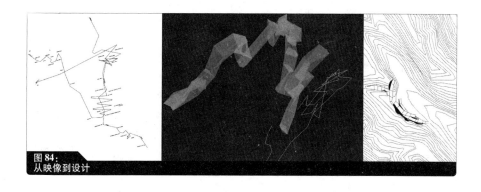

图84：
从映像到设计

你既可以从物质对象，又可以从非物质的现象中找到母题的灵感，并把它转化成你设计中的元素。其困难性取决于这次转变所提取的程度能否被准确地界定。如果你所选取的母题因为非常清楚的主旋律和联想而与众不同，你必须要确定，你的设计不会因为你直接应用该母题而显得过于简单乏味。同时，你还必须要确定，你的母题不会由于与其原始的母题失去关联而显得过于抽象。你的设计应当对母题的影响作出微妙的反应（见图81～图83）。这些联系有助于你学会如何利用参考性母题、抽象性和象征性元素。

映像

映像（mapping）是一种抄写的特殊形式，也是一种开发设计主题的特殊方法。它使得你的设计与未来建成后场地的特殊氛围建立起关联性，并作为你设计的基础。映像的实际程序可以包括拍照摄像、画速写、做模型等。然后，这些数据要通过某种特定逻辑和语言——一种特殊的注记系统——进行评估，重点在于空间氛围（如动线、光影、缺口、噪声等）。注记可以把空间氛围转译成典型的建筑结构（见图84）。这种对基地的强大而客观的分析能够使你对基地的研究更深入，更有效地理解和把握设计任务。

通过试验生成形态

除了以分析的途径着手设计，你也可以通过尝试从形态创造入手来取得初步的设计主题。你可以计算机上利用设计和视觉化软件来执

> 提示：
> "映像"一词源自地图（map）。地图关乎事实（facts），地图用图形表现存在于物质世界或者逻辑世界之中各元素的尺度、属性和相互关系。几乎所有的事物都可以被描述和记录：空间、星系、时间、历史、职业和思想。如果你对这个专题感兴趣，我们推荐你去读由罗杰·佛赛－唐（Roger Fawcett-Tang）和威廉·欧文（William Owen）所著的《映像》（Mapping）一书。该书2005年由RotoVision出版。

图85：
通过实验性方法创造出来的形态的建成实例

行这一任务，也可以利用模型方法（如石膏、蜡、橡皮泥等）。你可以使用三维扫描仪或其他设备把模型数字化。通过对传统材料在不同环境下的试验，你可以创造出无法预料的、创造性的形式。

一些可用的方法如下：

— 从揉皱了的纸团中获取形态（弗兰克·盖里研究方案时就是这么做的）；
— 通过使一个物体（比如一个易拉罐或者一个纸盒子）变形来创造复杂的几何形态；
— 利用磁力原理创造结构：如在桌面上排列铁屑或者铁钉；
— 在一个平板上用橡皮泥创造动感形态。

实验性方法可能一开始有一种偶然和随意的性质，然而，这种情况仅仅在一定程度上是的确如此。你会不断尝试并不断放弃各种想法，直到你认为最终得到了一个有潜力的结果。然后你会对此结果进行修正和优化。通过这种方式，实验成为了你创作过程中的催化剂。随后的设计过程从这第一步（作为基本的几何形态）开始。令人兴奋的是，无论你采用什么方法，你都可以确定你所创造的形态在功能和美学方面的潜力（见图85）。

结　语

尽管建筑设计的质量是通过一些特定的个性特征表现出来的，例如思维的连贯性、方法的多样性等，但对于设计质量的判断总是包含一些主观感知的方面。建筑的品质根本不能在客观的参数和尺度上被判断。评价常常包含着评价者主观的看法。人们的建筑品位的周期性变化也起一定的作用。20世纪80年代很现代的建筑到了今天会显得过时，而19世纪晚期欧洲鼎盛时期（Gründerzeit）的以抹灰饰面的建筑如今却非常流行，尽管现代主义者嘲笑其粗制滥造。无论一个设计是否追求永恒，或者追逐时尚，真正持久的设计是看其是否构思巧妙且被不折不扣地实现。设计过程的演进从一个初始的概念出发，直至设计逐渐成型。设计的目标是把设计中潜藏的原理和基本概念转换成设计方案，设计出细部，一直到施工阶段。

本书试图为设计者提供一些启示——必须注意的是，它只涉及了一个过程的初步阶段。书中所展现的各种方法也并非定论，仅仅是指出了处理设计问题的种种途径。设计不是简单地重复一定的程序和现存的模式，设计总是在各种条件下、从各种源泉获得启发、生成全新整体的创造性过程。本书中所分别探讨的形成初始设计概念的那些设计元素——文脉、功能、设计、材料、结构——必须在设计过程中互相作用，以创造一个影响和条件的复杂的系统。哪一种因素促成了第一个设计概念的生成？哪一种方法导致了设计概念的成熟？答案取决于每个人的经验和洞见。不但如此，建筑学课程为获得个人经验和洞见提供了一个宝贵的基础。设计不是可以被动地从书本和课堂上学习的，书本和课堂只能为学生个人发展提供一些刺激，起到催化剂的作用。设计是需要在操作中学习的——通过实践经验。我希望《设计概念》这本书能够为你的创造性设计提供某种刺激，能够鼓励你致力于去探索设计的各个方面。通过本书，我希望从长远来看，能对你探索设计之路有所帮助。这意味着，你必须不断地追问，勇敢地尝试，保持你的好奇心，在设计中发现创造的乐趣。

附 录
APPENDIX

参考文献
LITERATURE

设计基础

Leon Battista Alberti: *The ten Books of Architecture:* the 1755 Leoni Edition, New York 1986

Francis D.K. Ching: *Architecture. Form, Space and Order*, John Wiley & Sons, New York 1996

Le Corbusier: *The Modulor, Modulor 2*, Birkhäuser Verlag, Basel 2000

Roger Fawcett-Tang, William Owen: *Mapping*, RotoVision, Brighton 2005

Christian Gänshirt: *Werkzeuge für Ideen*, Birkhäuser Verlag, Basel 2007

Jeffrey Kipnis: InFormation/DeFormation, in *Arch+*, No. 131

Greg Lynn: Das Gefaltete, das Biegsame und das Geschmeidige, in *Arch+*, No. 131

Andrea Palladio: *The four Books on Architecture*, MIT Press, Cambridge 1997

Camillo Sitte: *City Planning according to artistic Principles*, Phaidon, Berlin 1965

Vitruvius: *Ten Books on Architecture*, Cambridge University Press, Cambridge 1999

建筑历史与理论

Otl Aicher: *The World as Design*, Ernst, 1994

Leonardo Benevolo: *The European City*, Blackwell, Oxford 1993

Le Corbusier: *Towards a new Architecture*, Dover Publications, London 1986

Siegfried Giedion: *Space, Time and Architecture: The Growth of a new Tradition*, Harvard University Press 2003

Hanno-Walter Kruft: *A History of architectural Theory from Vitruvius to the Present*, Zwemmer, London 1994

Robert Venturi: *Complexity and Contradiction in Architecture*, Little Brown & Co 1977

Robert Venturi, Denise Scott Brown, Steven Izenour: *Learning from Las Vegas*, MIT Press, Boston 1972

其他关于设计的文献：

Jürgen Adam, Katharina Hausmann, Frank Jüttner: *Entwurfsatlas Industriebau*, Birkhäuser Verlag, Basel 2004

Sophia und Stefan Behling: *Solar Power*, Prestel Publishing, Munich 2000

Mark Dudek: *Entwurfsatlas Schulen und Kindergärten*, Birkhäuser Verlag, Basel 2006

Roberto Gonzalo, Karl J. Habermann: *Energy-Efficient Architecture*, Birkhäuser Verlag, Basel 2006

Rainer Hascher, Simone Jeska, Birgit Klauck: *Office Buildings. A Design Manual*, Birkhäuser Verlag, Basel 2002

Paul von Naredi-Rainer: *Entwurfsatlas Museumsbau*, Birkhäuser Verlag, Basel 2004

Ernst Neufert: *Architect's Data*, Blackwell Science Ltd, London 2000

Friedericke Schneider (ed.): *Floor Plan Manual. Housing*, Birkhäuser Verlag, Basel 2004

图片来源

图 1，图 3，图 5（W. Wassef），图 12 左（R. Moneo），图 26 右（E. Heerich），图 30，图 34 左 + 中，图 35 全部，图 46，图 49 左（H. Guimard），图 70（H. Stubbins），图 71 右（T. Herzog），图 77（M. Campi, F. Pessin），图 80 Sebastian El khouli

图 2（Morger Degelo），图 46 中，图 75 左（HdM） FG ee，Hegger

图页 10（Le Corbusier），图 4、6、11、12 右，图 13，图 14，图 18，图 20（Botta, Palladio），图 23，图 24，图 26 中左、中右（Gaudí），图 27，图 29 右，图 33（Forster），图 36（Le Corbusier），图 39，图 41，图 47 全部，图 48 全部，图 49 中左 + 右（Gaudí），图 55，图 56 左（Mendelsohn），图 56 右（Le Corbusier），图 58，图 60 ~ 图 63，图 64 左（Mies van der Rohe），图 65 ~ 图 67，图 71 中左（Gaudí），图 71 中右（Le Corbusier），图 73，图 74，图 75 右，图 81 全部，图 85 中右（Domenig） Bert Bielefeld

图 7 左 Katrin Kühn/ Sonja Orzikowski

图 7 右，图 21 Barbara Gehrung

图 8，图 15，图 25，图 57，图 59 Isabella Skiba

图 9，图 19 Rahel Züger

图 7 中，图 10，图 16，图 17，图 26 左，图 34 右，图 40，图 83 Atelier 5

图 22 Joost Hartwig, Nikola Mahal

图 28，图 29 左，图 78 Annette Gref

图 31（Atelier 5）	Leonardo Bezzola
图 32（Brüning Klapp Rein）	Brüning Klapp Rein Architects
图 37，图 69，图 84（DGJ）	Hans Drexler
图 38（R. Serra）	Andrew Dunn
图 45（Le Corbusier）	FLC，ProLitteris，Zurich
图 49 中右（G. Strauven）	Peter Clericuzio
图 50，图 51（Frei Otto），图 53（Grimshaw），图 64 右（M. van der Rohe），图 71 左，图 72 左（Calatrava），图 72 中（Frei Otto）	Free pictures
图 52	Gerd Alberti, Uwe Kils
图 54	arp
图 56 中（Utzon）	Denn
图 68（F. Gehry）	Jon Sullivan
图 72 右（Nervi）	I have got the style
图 76（Behnisch）	Chris 73
图 79（P. Cook，C. Fournier）	Marion Schneider, Christoph Aistleitner
图 82（Zamp Kelp/Krauss/Brandlhuber）	Cordula
图 85 左（Z. Hadid）	Richard Wasenegger
图 85 中左（Coop Himmelblau）	Andreas Pöschek
图 85 右（Gehry）	Cacophony
图页 80	Martin Zeumer

所有图片均已注明出处并获得使用许可。如果发现有任何错误或疏漏之处，请与 Birkhäuser 出版社联系。